COMMUNICATION TECHNOLOGY, MEDIA POLICY AND NATIONAL DEVELOPMENT

By the Same Author

Communication Planning for Socio-Economic Development, edited alongwith Ranjit Singh

Media and Development, Themes in Communication and Extension, edited alongwith M. R. Dua

Third Revolution in Indian Perspective–Contemporary Issues and Themes in Communication

Rural Press–Problems and Prospects, edited alongwith Rajeshwar Dyal.

Media Policy and Nation-Building–Select Issues and Themes, with Vir Bala Aggarwal

National Media Policy, edited alongwith Rajeshwar Dyal

Media and Market Forces–Challenges and Opportunities–edited alongwith Rajeshwar Dyal

---

About the Author

V. S. Gupta, currently Professor at Haryana Agricultural University was awarded the UNDP fellowship for training in Europe and selected for the Journalist Fellowship Award for working in the areas of Environment and Development. He conceptualised and implemented as Course Director, the Post-Graduate Diploma Course in Agricultural Journalism through Distance Education and Refresher Course in Communication and Writing Skills, in the University. He is the Country Representative for the Asia Media Information and Communication Centre (AMIC), Singapore. Reckoned as a well known media commentator in India, Prof. Gupta is associated with half a dozen departments of Journalism and Mass Communication at the post-graduate level in the Indian universities.

COMMUNICATION TECHNOLOGY, MEDIA POLICY AND NATIONAL DEVELOPMENT

V. S. GUPTA

CONCEPT PUBLISHING COMPANY, NEW DELHI - 110 059

ISBN 81-7022-745-3

First Published 1999

Published and Printed by
Ashok Kumar Mittal
Concept Publishing Company
A/15-16, Commercial Block, Mohan Garden,
NEW DELHI - 110 059 (India)

Phones : 5648039, 5649024
Fax No. : 91-11-5648053

PREFACE

This book has been much longer in the making than was planned. The media landscape in the country has been undergoing rapid changes–more so during the last two years. Several media policy issues and regulatory measures are being implemented post haste, these are being undone equally hastily. The changes in the major policy enactment–Prasar Bharati Bill–being brought about through ordinance, are being undone through the lapse of ordinance. While all this is happening, the technological advances are overtaking us equally speedily. All these developments–exciting as these are–involved revision of several portions in the relevant chapters.

Apart from the extensive revisions that the changing policy perspectives entailed, my own propensity to keep on updating the information resulted in deferring deadlines over the past two years. The result : Deadlines were deferred; deferred deadlines became distant and hazy; the phrase with which I got in love and wished to implement so vigorously with the course participants in our university–there cant' be a deadline, beyond a deadline, became the worst casualty. In any case, now the book is finished, I should not mull over the whole thing much too much.

The rationale of the book has been explained in the Introduction and Overview. Briefly it is : to take note of and analyse the trends and developments in the media scenario in the country–India's progress towards becoming an Information Society, Issues in the National Media Policy and the strategy to harness the potential of media for National Development. The whole undertaking has been indeed exciting and rewarding.

ACKNOWLEDGEMENTS

My deepest debt of gratitude goes to AMIC and its Secretary General, Mr. Vijay Menon. My close association with AMIC and the inspiration and support provided to me by Mr. Menon over the years, as a member and also as a participant in several of its academic programmes, seminars and conferences helped me build insight and perspectives on several media-related issues.

My association with the Friedrich Ebert Foundation has been equally rewarding. I must, therefore, place on record my gratitude to the Foundation for the generous supply of media publications; to Dr. Heinz Bongartz and Mr. Horst Mund, the former and the present Resident Representatives; Dr. Peter Hering, Head, Media Programmes, Asia; Senior Media Advisor, Mr. Rajeshwar Dyal, and Mr. Kabir Seth who would always provide instant solution to all the problems. I am hopeful this rewarding association will continue in times to come.

I am thankful to several of my colleagues, associated with me on different occasions, Mr. R. V. Rajan, Committee Member, AMIC, Prof. M. R. Dua, IIMC; Vir Bala Aggarwal, from the Department of Journalism and Mass Communication, Himachal Pradesh University, Shimla; Usha Chander from IGNOU, Sucharita S. Eashwar, *Voices,* Bangalore; and Indira Bajaj, Deputy Librarian from my university, who was always prepared to walk an extra mile in helping me out.

To the publisher–Mr. Ashok Kumar Mittal–a friend, whose personal and professional advice I could always count on; and who patiently put up with deferred deadlines time and again; my heartful thanks.

My efforts in the past in bringing out a number of publications on media-related subjects have been acknowledged and recognised by a large cross-section of readership throughout the country. An appreciative readership–one could not ask for more. It is with the same hope in my heart, and prayer on my lips, that I come to the end of my labours.

15 May 1998 V. S. GUPTA

CONTENTS

1

INTRODUCTION AND OVERVIEW

The exponential growth in the media industry particularly in electronic and satellite communication, and rapid advances in communication technology, witnessed during the recent years in India, has changed the media scenario as never before. The proliferation of new information and media technologies has brought about fundamental changes in several spheres of human activity–life styles, entertainment, work culture, modes of communication and has also influenced social and cultural values in the society. An other issue confronting us today is the increasing compulsion of the media to submit completely to the market forces.

In the wake of these developments, the focus has shifted from development support communication to computerization and digitization, from rural communication to satellite transmission, from information to infotainment. Advances in communication have led to fresh thinking on two aspects–to strengthen regulatory mechanism and media monitors on the one hand and explore the informational and education potential of the media. The contemporary media scenario bristles with a number of significant issues–convergence of computing, telecommunication and broadcasting technologies resulting in the ushering of Information Revolution in India and its social implications; informatization of society and commodification of information are the newly emerging issues.

For the past several years formulation of National Media Policy has been one of the pre-occupations with the Ministry of Information and Broadcasting. The debate on National Media Policy has been enriched by the deliberations of a number of committees appointed over the years by the Government of India to study various aspects of media policy : autonomy of electronic media, software production, distribution of airtime, cable television, etc.

The relevance and usefulness of media technology and media policy, in the final analysis, has to be tested at the touchstone of national development. The extent media can be harnessed to fulfil various objectives of nation building such as National Unity, Rural Development, Literacy and Distance Education, Environmental Protection and Sustainable Development, Women Empowerment, etc. has been discussed in the chapter–Media Practice and National Development.

A modest effort has been made in the book to explore the contemporary trends and developments in media and communication technologies, the way they are impacting the society as also the extent to which these technologies could be harnessed in the cause of nation building.

Rationale and Scope

The book has been divided in three themetic sections, the section on Communication Technology comprising two chapters–India's Information Revolution and New Information and Communication Technologies and Social Imperatives, looks at various issues associated with India's Information Revolution and its social implications; the second section on National Media Policy looks at the objectives of media policy : Electronic Media Policy, Prasar Bharati Act, Broadcasting Bill, Press Council of India and Entry of Foreign Press in the country. The third section covers the core issues of nation building–shifting paradigms of development communication, media strategy for rural development, media practice with reference to national unity, literacy and education, protection of environment and sustainable development, and women empowerment. Finally, the concluding chapter looks at some other emerging issues and aspects presenting confronting the media, and in a way updates the developments.

INDIA'S INFORMATION REVOLUTION AND NICTs

In their book, *India's Information Revolution,* (1989) Arvind Singhal and Everett Rogers, made one of the first efforts to take note of societal developments taking place in India as it moved towards becoming an information society. During the intervening years, the progress in the direction has been quite perceptible. The book refers to 'Silicon Valley fever', eventually happening in certain places in

India, and mentions Bangalore, Pune and Chandigarh to be the most likely places for such high technology agglomerations to occur. Today, however, Hyderabad has come up very fast as the 'Silicon Valley of India'. With Microsoft selecting Hyderabad for its first software development centre outside United States, it is all set to become the information technology hub of the country. As a part of its multipronged strategy to turn Hyderabad into 'Cyberabad', the state government is planning a 'Hitech City'—Hyderabad Information Technology and Engineering Consultancy City. This will not only make Hyderabad the hub of it but also provide local talent an exposure to the cutting edge technology and latest projects.

India a Software Power

That India's progress towards becoming an Information Society is bound to be fast, is quite clear from the priority that IT has received in the National Agenda for Governance formulated by the present Government. The National Agenda says : A new revolution is sweeping the globe–that of Information Technology. We believe that Information Technology is an important vehicle for future development. We are committed to ensuring that India is a full and equal partner in its global progress; that India can be a software super power. We will, for that purpose unveil a National Informatics Policy with short, medium and long term perspectives.

That the new Government is indeed earnest in its effort to promote information technology in the country is evident from the fact that even the 5-point economic agenda announced by the Prime Minister includes an item on making India a software super power.

This section explores the progress of tele-communication, touches upon the National Telcom Policy and the role assigned to the tele-communication industry in the Approach Paper to the Ninth Five Year Plan. Some other subjects discussed are : an emerging new breed of information entrepreneurs, India's progress from SITE to INSAT and GRAMSAT-satellites tailored to cater to the basic needs of rural areas in the country.

The one disquieting aspect of Information Revolution is : To what extent a country like India beset with stark internal contradictions- industrialised cities and marginally subsistent villages, jet planes and bullock carts, satellite technology and lack of clean drinking water, can make the transition to Information Society more equitable than is possible under the present social and economic dispensation.

The social impact of new information and communication technologies continues to nag the author even in the second chapter of the section on Communication Technologies. Several aspects of Information Revolution are further purveyed in the chapter, New Information and Communication Technologies and Social Imperatives. Information is fast becoming a commodity, leading to commodification of Information. The background and several formulations about Information Society, in themselves, constitutes a fascinating subject of study. The increasing references to multimedia and information superhighways, not only as key carriers and processors of information content, but as resources for further economic development, thus creating elaborate infrastructures has given rise to glowing euphoria largely unwarranted.

While the present author does not by any means belong to the Luddite section, which is characterised by opposition to technological innovation, the nagging realisation about affording primacy to technology as a means of redressing people's problems continues unabated. In this context the views of one of the foremost communication thinkers and researchers of the Third World, Neville D. Jayaweera with reference to claims of efficacy of communication satellites to development and reconsideration of these claims continue to be as relevant today as when these were articulated in 1982. The message, the views of Jayaweera reproduced briefly in the tabulated form, puts-forth is rather simple : Technological devices cannot be a substitute for political action and initiative. Equity concerns, in the context of impact of new communication technologies, would continue to dominate the equality-inequality question.

Referring to the difficulties attending the study of information society, Professor William J. Martin, Head of the Department of Information Management, Royal Melbourne Institute of Technology and author of *The Global Information Society* says : One of the weaknesses attending the study of information society todate has been that much of it has entailed reaction to various forecaste or predictions by futurologists. Quite apart from the, outrageous claims appearing in some of these forecasts, their less than impressive record of success has scarcely added to the credibility of notions of an information society. Tom Forester has attributed the lack of success of these forecasters to the fact that they were driven by Utopian visions of technological change rather than by practicalities and hard realities.

It would, therefore, be essential for the planners and policy-makers not to be swayed by technological determinism, the concept that technology is the main cause of social change in a society. Any extreme position in this regard will aggravate the technological dilemma. It would be equally essential to evaluate alternative choices to steer the society to its destined goal.

NATIONAL MEDIA POLICY

Formulating a National Media Policy for a country, as vast as India, a democratic nation with regional, cultural and linguistic pluralities, would have been a complicated task at best of times. More so now at a time when new technological imperatives like advent of satellite communication, transnational electronic printing facility, information superhighway and convergence of communication and media technologies have made implementation and regulatory aspects of such a policy a challenging one.

Whenever the issue of formulating the National Media Policy came up, it was considered axiomatic that the objectives of such a policy be laid down. Some experts even questioned the rationale of such a policy. The former chairman of the Press Council of India, A. N. Sen opined that it may not be wise to seek to lay down any firm National Media Policy, as the effective implementation of the policy may create more problems than it is likely to solve.

The views of media critic D. R. Mankekar, analyst Pran Chopra and media educator, M. R. Dua; as well as the objectives laid down by Consultative Committee of the Ministry of Information and Broadcasting in its report submitted in March 1996, have been listed. Several committees, working and study groups came up with their recommendations on different aspects of the working of media. The salient features of these reports constitute valuable source material for this section.

This chapter, initially written up in June 1997, had to be revised several times, the latest revision was necessitated around 15 April 1998 in order to incorporate the current thinking on the subject as enunciated in the National Agenda for Governance.

Some of the steps being contemplated presently are : allowing the Presidential Ordinance regarding Prasar Bharati Bill to lapse which would enable the government to reconstitute the Prasar Bharati Board as also to revive the Parliamentary Committee, as envisaged under Article 13(1) of the Act to "oversee that the Corporation discharges its functions in accordance with the provisions of the Act". Incidentally the present author, as the AMIC Country Representative coordinated a series of five seminars during 1995 on

National Media Policy in different parts of the country. The seminars, sponsored by Friedrich Ebert Stiftung, enabled a large number of media professionals and academics to present their views on different aspects of National Media Policy and the Prasar Bharati Act. The view that emerged from the seminars was that the Parliamentary Committee (which is sought to be revived now) would invite definite political interference in the working of Prasar Bharati Board. However, another guiding principle in the National Agenda for Governance which moots that the foreign equity should not exceed 20 per cent in private television broadcasting (as against 49 per cent in the Broadcast Bill, introduced by the United Front Government) would be a welcome step in that it would ensure protection of Indian interests. The above provision implies that the Indian investors can have upto 80 per cent stake, and therefore places Indian channels and investors in a highly advantageous position. Since media are a national resource of great educational and cultural importance it is natural that Indian channels and investors have a position of advantage. Some other important aspects of the National Media Policy being proposed presently are : The Broadcasting Authority of India as well the code incorporating the guidelines for controlling satellite invasion, uplinking facilities from the Indian soil and ensuring proper implementation and monitoring of the guidelines. The value of the overview of this section has been enhanced in that it has enabled the author to update and incorporate the latest thinking on the subject.

MEDIA AND NATIONAL DEVELOPMENT

Emerging Paradigms and Concerns

The core section of book takes up issues which concern social scientists, planners and policy-makers, development and NGO sector, and of course academics in the discipline of media and communication. The section takes note of the shifting paradigms of development–from the GNP approach right down to sustainable development and now human rights approach, and emerging concerns–why the benefits of planned development have not percolated to the grass root level, strategies for rural development and media practice for strengthening concerns like National Unity, Environmental Protection and Women Empowerment; and of course the role of media technologies in aid of Distance Education.

The growth of Development concept from GNP and per capita

income to human aspects of development linked to quality of life; to becoming a part of Declaration on Rights to Development adopted by UN General Assembly on 4 December 1986, to Sustainable Development in the wake of emerging crisis of environmental deterioration; offers a fascinating debate and discourse on the emerging paradigms of the theories of economics, social change and communication.

The Dag Hammarskjold Foundation Report–*What Now? An other Development* (Uppsala, 1975) gave new direction to the concept of Development. And, of course, as pointed out in the chapter on Development Communication, Development would mean different things to different people in different countries depending upon the national priorities; one-child family in China, decreased pesticide use in Philippines, ORT, clean drinking water and rural roads in India. The UNDP Human Development Reports since 1993 have gone a long way in redefining the concept of Development.

And an absolute delight was going through *Seminar* (March 1997) wherein it was pointed out, regret and rethinking on the part of Professor Tinbergen, who propounded the concept of GNP as a measure of progress, because it was steering the society with a wrong compass. Truly the days of redefining the old concepts, retooling with the new ones and reformulating the premises are with us. And not a day soon.

Even the apex international bodies like World Bank are giving a second look to their aid programmes. Unthinkable a few years ago, leading religious leaders from around the globe reached a landmark accord with the World Bank which aims to put moral and social values back into the financial giants' balance sheets. This rethink was possibly due to a shift in the bank's understanding of development to include cultural, ethical and spiritual issues. Reporting the path breaking initiative, *The Pioneer* (3 April 1998) says : up until now, the World Bank has ignored the importance of cultural, ethical and spiritual values in development project because it has focused on economic values alone. Non-governmental development agencies have attacked many of Bank's policies, including its soft loans at lower rates for developing countries. This has sunk the developing world into a spiral of escalating debt, conditional on economic policies not tuned to the needs, cultures and existing structures in the recipient countries. The World Bank loans are usually made directly to the governments and into large scale projects such as Narmada dam in Central India which displaced thousands of people and caused extensive environmental damage.

Let us hope the dialogue was not a "one-off firework which blazes for a moment and then dies".

There has been growing evidence that media can play significant role in enhancing public activism and in drawing attention to the deprivations of the socially and economically backward sections of the society.

As a part of abundant caution it has been stressed that Development Communication should not be considered a part of public relations for development departments. And of course participatory approach–an essential part of Development Communication–needs more social learning, risk taking and coordination skills, besides political commitment.

The chapter, Media Strategy for Rural Development, highlights the localness of approach and dissemination of culturally appropriate and user friendly technology. Rural Press–its effectiveness and potential, as a link between policy makers, development bureaucracy and farmers; and also as second line of conversion of scientific knowledge into information for utilization in the field situation has been discussed as also the need for the rural press to design an appropriate editorial formula that is attractive enough for the readership in the community. Constraints in reporting rural development notwithstanding, press in Kerala has been in the forefront in rendering yoeman service to the farming community in the state.

The rural broadcasting programmes of All India Radio, through F&H units, have been quite effective and popular in disseminating agricultural technology information to the farmers but, as the veteran broadcaster Duvpesh Chandra Bhaumik points out we can't afford to ignore peoples' point of view and have to change our mind set of two-step model. The commendable initiative taken by the leading media NGO in the country, *Voices,* in conceptualising and promoting community radio needs to be strengthened further.

Several research studies conducted to evaluate the effectiveness of the Krishi Darshan Programme have concluded that TV is perceived to be primarily a source of entertainment. The Doordarshan study conducted by its Audience Research Unit in January-March 1993 pointed out that even in villages number of persons watching the feature films and chitrahar were nine from a single TV set as against two persons watching programmes on agriculture. There is, therefore, an urgent need to revamp the entire set up of Krishi Darshan, in case television wishes to contribute effectively towards rural development efforts.

The chapter, Media Practice and National Development discusses role of media in fostering National Unity, Literacy, Educational Technology and Distance Education, Environmental Reporting for Sustainable Development and Women Empowerment–Role of Media; Beijing and Beyond. Developing the theme of National Unity and elaborating the concepts like Nationalism, National Identity, Sub-nationalism was a delightful and inspiring experience. Instances from China where media could be harnessed for enhanced national awareness, and Nigeria where media are viewed as national unifier, are particularly noteworthy. The Indian experience has been different. Whereas the Electronic Media–so far under the control of Government–are mandated to uphold the unity of the country, it has been widely felt in practice that Delhi-centric view has been projected to the neglect of other parts of the country. There seems to be some validity in the perception that at times enough sensitivity has not always been shown to the cultures, customs and languages from states from the Southern and North Eastern parts of the country.

We will do well to learn from the Indonesian experience where 'what has been viewed as national integration from the centre has often been seen as imperialism from the periphery'.

The role of the press—and particularly that of the vernacular press—has not always been edifying in times of national crisis; and in several countries it has encouraged sectarian, ethnic and linguistic passions. In such cases media monitors, regulatory bodies and professional associations have a role to play.

Education and media have to be harnessed to bring about changes in attitudes and values—so essential for the task of nation building. Whereas media certainly have the potential to aid the educational system of the country the success depends upon ensuring that media technology fulfills the needs of learners, teachers and society. Media are particularly suited as adjuncts to Distance Education and Open Learning System. The fact remains that Open and Distance Learning System would not have created a viable and lasting nich for itself, within the overall traditional system of classroom teaching and university education, but for the extensive use of media and communication technologies. Communication not only generates an educational environment, it itself becomes a vehicle for education.

Media technology is being increasingly harnessed to expand educational opportunities. China is one of the pioneering countries in the world to have exploited television to meet its educational needs by establishing Central Radio and Television University. Tele-education programmes, the use of educational teleconferencing for teaching and training purposes by IGNOU and National Open School are some other instances in this context. Transnational teaching and virtual classroom have become possible due to the convergence of computing, satellite and telecommunication technologies.

Since the 1972 Stockholm UN Conference on Environment, issues concerning environment and development have received sustained media attention. Environmental reporting focuses virtually on contemporary crisis of survival. The media coverage of grass root movements throughout the country has certainly enhanced the awareness level of environmental issues. Since environmental problems are assuming global dimensions and impinge upon decisions taken at multi-lateral levels, the interest generated ranges from the local issues of urban sanitation and deforestation to global warming and green house effect. Environmental issues have also generated 'rethink' about the aid and funding procedures of the World Bank, construction of big dams and other environmental projects.

Judicial activism and concern expressed by the Supreme Court in several cases of environmental pollution, deforestation and deep fishing in coastal areas, alongwith public activism of voluntary agencies has moved environmental reporting up in the priority category.

The unprecedented media hype associated with the Fourth UN Conference on Women (Beijing, 1995) once again brought women issues on the world agenda. Awareness apart, issues concerning women empowerment have been taken up by the print media, particularly the regional press, as a subject for systematic reporting. Since several issues like population and development have close linkages with the problems women face throughout the world, activities and decisions of international organisations, national associations of women and government institutions and departments provide abundant opportunities for media reporting. This section has, therefore, also discussed recommendations of the Beijing Conference in some detail for wider dissemination, opinion building and to the extent possible, implementation by those at the decision making level in media organisations.

The book addresses some of the fundamental issues concerning the role and impact of new information and communication technologies, the on-going developments in National Media Policy and those aspects of nation building where media made distinct contribution in the past, and have the potential to make in the times to come.

REFERENCES

Asian Mass Communication Research and Information Centre (1996). *Opening Windows, Issues in Communication,* Singapore.

Caring for the Future (1996). *The Independent Communication on Population and Quality of Life.* Oxford University Press, New York.

Dale Penny (1998) A more humane World Bank, *The Pioneer,* 3 April.

Martin, William J. (1995). *The Global Information Society,* Aslib Gower, Hampshire.

Mehta, Dewang (1998). Future Perfect, *The Pioneer,* 30 March.

Narayanan, V. N. and Sabharwal, Jyoti (1997). *India at 50, Bliss of Hope and Burden of Reality.* Sterling Publishers, New Delhi.

Singhal, Arvind and Rogers, Everett, M. (1989). India's Information Revolution, Sage, New Delhi.

Voices (1997). *Media Policy and Community Radio,* Bangalore.

2

INDIA'S INFORMATION REVOLUTION

The Information Revolution in India got off to a slow start–feeble and haltingly it made its beginnings in 1970s. The contours of Information Revolution began to emerge when computers were initially adopted by government departments to facilitate government procedures and expedite information processing. Several industries such as railways and airlines initiated computerisation of reservation services. It was only after the establishment of Computer Maintenance Corporation (CMC) and introduction of micro-computers in India the large potential for computer market was envisaged in the country. The introduction of colour transmission in 1982 in the wake of the Asiad, and the approval of the Special Plan for the expansion of the television network in 1983 led to the boom in broadcasting facilities and private ownership of television receivers. The liberal policies of Prime Minister Rajiv Gandhi during 1984-89 ushered in telecommunication and computer revolution in the country.

Since 1991, and more so during the recent years several trends–rapid pace of diffusion of technologies, liberalisation of economy, sustained high growth in telecommunication sector and exponential growth of cable and satellite television–have gathered momentum at a speed, which newspaper articles have frequently termed the current situation as, 'super information explosion', and 'multi-media revolution'.

As of now, a quiet revolution is taking place in the Indian business and the corporate world. Business corporations, public sector undertakings, banks and financial institutions, even universities, research and academic institutions have taken to computer networking. Several national computer networks like NICNET, ERNET and EDNET have been operational for several years. The value-added telecom products and services are flooding the metropolitan and urban areas. India has certainly taken first steps towards the Information Revolution.

Telecommunication

The telecommunication sector has emerged as one of the fast growing sectors in India and is reckoned as having immense potential for future growth. During the previous Five-Year plans telecommunications made considerable strides both in quantity and quality; however, its growth in India as compared to the developments in the world telecommunications can at best be termed as modest. Even by the standards of many developing countries, India is far behind in providing telecom services to the people. According to the latest projections, the telephone density in the country at 1.6 per hundred is less than that in Egypt (5.1), Malaysia (7.37) or even Thailand (6). Large population and the huge size of the country have adversely affected the telephone density in India inspite of the fact that telecom network has recorded considerable growth in the recent past, which has been the outcome of the liberal policies of Prime Minister Rajiv Gandhi.

With the objective of enhancing subscribers satisfaction level by improving quality of service and increasing accessibility of telecom facilities both in urban and rural areas in a time bound way Telecom Mission, popularly known as, Mission : Better Communication, was launched in April 1986. To give the nation a continuous improving telecom service, areas of thrust like quality of service, accessibility in rural and urban areas, modernisation of telegraph service, penetration of telecom service in rural, hilly and tribal areas and a major programme of setting up of national digital network were selected, clear cut objectives and specific targets were fixed, strategic and operational plans were drawn to achieve the Mission's goals. The time bound programme resulted in substantial improvement in various services. As on 31 March 1995, there were 87,543 STD/ISD PCOs, 14,3002 local PCOs, 2010 Highway pts, and 53,228 Telex connections in the country.

There has been a change in perception regarding the importance of telecommunication services which are no longer considered elite services. The simplistic thinking : A man can live without a telephone, but can he survive without drinking water, earlier came in the way of the development of the telecommunication sector. Now the enormous planning and logistics support that an efficient telecom sector can provide to all sectors of economy, and the resultant gain in productivity, is being duly taken into account in determining priorities.

Telecommunications is now taken as a prime support service needed for rapid growth and modernisation of all sectors of the economy which needs to be promoted and developed as a basic infrastructural input.

The Eighth Plan witnessed a sustained high growth in the provision of basic telecom services. But the demand has far exceeded the supply resulting in large unsatisfied demand. At present, about 2.2 million people are waiting for this basic facility. During the Ninth Plan, telephone demand is expected to grow from about 18 million lines in the first year to about 36 million lines in the terminal year i.e., 2001-2002. To achieve the goal of providing telephones on demand, about 21.5 million new connections would have to be provided.

National Telecom Policy

The National Telecom Policy 1994, has outlined the basic policy framework for the future development of the telecom sector in the country. The stated objectives of the policy are : to ensure telephones on demand, to achieve universal coverage, to ensure world standard quality of services to consumers, emergence of India as a major manufacturing base for major telecom equipment and protecting defence and security interest of the country. The Ninth Plan will aim to achieve these basic goals in a time bound manner.

Realising the importance and potential of telecommunication industry in the economy of the country, the Approach Paper to the Ninth Five Year Plan (1997-2002) says : Telecom Policy has envisaged that India should emerge as a major manufacturing base and major exporter of telecom equipment. Manufacturing capacity of the indigenous industry is small in relation to the other major operators in the world and export constitutes a small proportion of the total production. Development of Indian multi-nationals in this area will be our goal in the Ninth Plan. Non-availability of latest technology, poor R&D base, non-adoption of exports as a strategy of growth and limited access to international financial markets for cheap funds are among the major constraints in the achievement of this goal. The necessary policy initiatives will be taken during the Ninth Plan to remove the bottlenecks. This would include encouraging joint ventures, rationalisation of custom and import duties on inputs and development of a strong industry sponsored R&D base.

During the Eighth Plan, public sector investment for expansion of telecom network for basic services was primarily financed by internal resources and market borrowings budgetary support constituting less than 1 per cent of the total plan outlay. This approach for financing public sector plan would be continued during the Ninth Plan. Though the private sector investment is expected in a big way, public sector will continue to play a dominant role in the provision of telecom services. Internal resources alone would be grossly inadequate to meet the increased requirements. Greater reliance will be placed on other sources of finance like raising funds directly from the public in the form of equity capital, market borrowing both from the domestic as well as international markets and through other means.

C-DOT

The rapid expansion of telecom services in the country has been greatly facilitated by the apex R&D institution—Centre for Development of Telematics (C-DOT). C-DOT was established by the Government of India in 1984, with a total budget of $30 million dollars for three years. It was mandated to design an indigenous telecommunications digital switching system. Created by Sam Pitroda, largely credited with revolutionizing telecom services in India, C-DOT successfully developed the technology for electronic EPABX systems and a 128-line rural telephone exchange. Prime Minister Rajiv Gandhi appointed in 1987 Pitroda Principal Advisor to oversee the five key technological missions; telecommunications, immunization, drinking water, literacy and oilseeds development.

C-DOT technologies have been exported to a large number of countries in Asia and Africa. The largest number of lines outside India are presently operating in Vietnam and Bangladesh.

The Department of Telecommunication has decided to franchise value added services to the Indian registered companies on non-exclusive basis. The value added services identified include : Radio Paging, Cellular Mobile Telephone Services; Electronic Mail, Voice Mail, Audiotex service, Videotax service and Video Conference; Bulletin Board Service.

Computer Revolution

Computer industry constitutes one of the fastest growing sectors in the country. From 1984 to 1988, the number of computers increased ten fold, the computer industry's revenues increased four fold and computer software exports increased five folds. This high-technology development in India is seriously starting to get underway. *India's Information Revolution* (Singhal and Rogers, 1989, p. 155). And there has been no looking back. The production of computers during 1994 was Rs. 13,000 million compared to Rs. 8,200 million in 1990. The industry has grown both in terms of physical output as well as range of products introduced in the market. The production of mini and micro computers based on Intel Pentium Processor has stabilised. Almost the entire demand for floppy disk drives, dot matrix printers, CRT terminals, key-boards and live printers is met from indigenous production.

Industrial production statistics apart, in India we witness offices proliferated with computers connected to each other with messages criss crossing over the wires using electronic mail. During the outbreak of plague in 1995, doctors in India effectively used Internet to communicate with their counter parts. A heartening development indeed realised by networking technology.

Networks have become popular in India especially during the last five years. In India, the most popular networking applications which have helped the common man are the railway reservation system, airline reservation system and the now evolving automated banking systems. World-wide networking by banks abroad, has enabled them to transfer funds across the globe within seconds or minutes as against days or weeks in the past. In the wake of liberalisation in the country and increasing competition from the foreign banks, the public sector banks have to gear up for the competition by introducing large scale automation.

NICNET is the most visible, the most ambitious computer based network of the government. The nationwide information network, NICNET, is a satellite-based information service developed for its operations at the central, state and district levels. The network helps decision-making for social, economic and development planning, implementation and project monitoring. It represents an advanced level institutional application of information technology to planning and provides a model of a centrally planned approach to information services.

Several other networks have also been pressed into service

thus widening the base for accessing remote information and communication through electronic mail, computer conferencing, teletext, facsimile, etc. The two important research and cooperative computer networks which India is making use are : CARINET, a network among development related organizations around the world (including Australia, Canada, Japan, Philippines, Singapore) providing information related to agriculture, health education, human rights, disaster relief, etc.

ERNET, the Education and Research Network, connects India's prestigious engineering schools, the five Indian Institutes of Technology (IITs) and the Indian Institute of Science in Bangalore, besides many other educational institutions. ERNET has over 250 nodes in the country and helps break hierarchical barriers by providing younger scientists direct access to information. Educational uses of networks include active exchange of information between researchers in India and abroad.

Internet, the most spectacular example of global computer networking, is opening up new highways of opportunities in India. It was only in late 1995 since the toll gates to the Internet were opened, yet transnational corporations, business houses, Corporate India, took to Internet swiftly making Networking the hot new sunrise industry in the Indian Infotech. There is also a growing realisation by the Indian Corporates that business today depends on managing information i. e. transmitting and receiving information at the right time, place and form. Corporate India has in the right earnest begun to tap the bewildering potential of computer networks, and with increasing number of people going online, the country is poised to join an international community of info-surfers.

In a feature on Technology–Internet, *India Today,* in its issue of 30 June 1996, describes how information entrepreneurs, an emerging new breed of professional designers, are offering web sites and home pages to the Indian companies. With the installed base of PCs in India now in excess of one million, networking is just taking off. However, with only an estimated 30 per cent of computer linked to networks, in both the public and private sector, India has a lot of catching up to do, as compared with countries like China, Malaysia, Taiwan and other Asian tigers, where more than 50 per cent of computers are networked. China has been leap frogging into the networking arena for the last three to four years, whereas in India the long hurdle race has only just begun. But begun it has, and is already gathering speed.

Satellites

Satellite communication in India, an upcoming important field, owes its origin to the establishment of the Indian Space Research Organisation (ISRO) in 1962. The UNESCO Panel of Consultants on space communications in 1966 suggested taking up a pilot satellite project, both for testing techniques and demonstrating the effectiveness of tele-communication satellites while serving the priority needs of the area through major contribution to development, which led to the well known SITE programme of 1975-76. The SITE was primarily undertaken to experiment with television through satellite communication to broadcast special programmes in 2400 villages. The SITE aroused wide international interest as India was the only developing country in the region to undertake an experiment using space age technology to bring its millions into the mainstream of national life.

From SITE to INSAT

The idea of an INSAT (Indian National Satellite System) had been first proposed in the late 1960's. The common plank on which various concepts of the system rested was that the potential of the satellite medium can be drawn upon for rapid expansion of TV on a nation wide basis for education, national integration and development. INSAT is a multi-purpose operational satellite system for domestic telecommunication, meteorological observations and data relay and direct satellite television broadcasting. It is a joint venture of Department of Telecommunications, India Meteorological Department, All India Radio and Doordarshan.

Innovative uses of INSAT system have already extended for the establishment of rural telegraphy in the north eastern part of the country and National Informatics Centres to provide nationwide data, communication links between district and state capitals, library networking on a large scale has been initiated. A channel on the INSAT has been now dedicated for interactive education.

The most spectacular impact of INSAT has been in the rapid expansion of TV which with 715 TV transmitters and direct reception sets, now provides access to over 80 per cent of India's population. Rapid increase in the regional services, use of portable earth stations and satellite news gathering vehicles, utilisation of satellite media for university education and nationwide radio networking has enabled even remote areas to become part of mainstream of the nation. The

Press Trust of India has implemented a system to provide its news and information services at high speed and increased volume by utilising the broadcast facilities of INSAT satellite.

GRAMSAT : Considering that eradication of illiteracy in the rural areas is of utmost importance so as to achieve all round development through ushering a new green revolution and assimilation of scientific ideas and attitudes into the cultural fabric of the Indian society, evolution of a new concept of dedicated GRAMSAT satellites, specially tailored to meet the basic requirements of rural areas has emerged. The GRAMSAT will also help in providing continuing education for special groups to ensure that their technical skills are periodically updated to overcome technological obsolescence.

With the rapid expansion of satellite and telecommunication services in India, a number of organisations in the public and corporate sectors, and companies are planning to use satellite communication networks for corporate communication or maintaining computer networks. *The Hindu,* a national newspaper published from Madras, has the distinction of being one of the few newspapers in the world which has made its editorial room paperless. The INSAT-1B satellite is used by the computers in the newspaper offices for bringing out a facsimile Delhi edition daily.

With Intelsat, leasing a portion of capacity of its proposed satellite 2E, India has entered the global space market in a big way. Intelsat is the international intergovernmental cooperative organisation providing telecommunication services for different regions of the world. The media world is unanimous that alliance with Intelsat will catapult India as one of the key satellite TV forces in the region. After the debacle of APSTE-2, this agreement, the first of its kind that the 135 nation Intelsat has signed with an outside agency, is bound to change the TV entertainment scenario in the country.

India's entry into space and boom in satellite capacity over the Indian skies will mean more satellite TV channels in major regional languages and better entertainment and news programmes. With increasing possibility of private entrepreneurs and industry entering the field, to create user-oriented information systems and products, the potential uses of space for information and communication are wide open.

Doordarshan is one of the few television organisations which uses domestic satellites for telecast of national and regional services. The successful launching of INSAT-2B earlier made it possible for

Doordarshan to telecast additional TV programmes through domestic satellite; with five transponders on the satellite widening the networks horizon through five new channels.

With the launching of Asia Sat 1 in April 1990–the major broadcast satellite agency over the Indian airspace–a cable TV Revolution has set in. Asia Sat 1's footprints cover most of Asia and parts of the Pacific. STAR TV launched the satellite television service through Asia Sat by the end of 1990. The number of STAR TV homes all over Asia increased rapidly.

Apart from Doordarshan programmes uplinked to the INSAT-2A satellite and the STAR to ASIA Sat, some other networks including CNN and Asia net uplinked to Arab Sat and Ekran also have contributed to the media hype that the country is currently witnessing.

The trail blazing efforts of STAR TV and intense competition to reach for the Asian skies, the satellite industry's revenues are expected to be in the range of $ 30 billion by the year 2000. According to Deodhar, Chairman of the Doordarshan's Airtime Committee : Space over Asia is becoming prime property. The booming economies of many Asian nations like India, China, Indonesia and Malaysia and their prosperous and demanding middle class have fueled a very high degree of interest in satellite communication services. The boom in satellite capacity over Indian skies will mean more satellite TV channels in all major languages, more entertainment, more news.

India's Information Revolution–Social Implications

A very large percentage of the population, living in small towns and villages, is not touched by the Information Revolution, raising several socio-cultural and economic issues, which have to be reckoned with. The extent to which beneficent aspects, which some of these technologies claim for society, could find application in nation-building activities, need closer examination. It would, therefore, be important to go beyond the hype and euphoria, as also the state of optimism generated by the expanding entertainment industry and information technology, which are essentially led by market-driven forces. What are the implications of the emerging Information Society for the common man and a lay citizen in the country, what do we do about the yawning gap, increasing by the day, between the information-rich and the information-poor and what do these societal changes portend for the future.

During the World's biggest telecommunications exhibition organised by the ITU in Geneva in October 1983, the Chairman of the world's largest operator of telecommunication, Charles L. Brown made some very pertinent and interesting remarks on the subject matter of the seminar . . . a revolution is taking place in communications technology. Innovation in telecommunications technology is proceeding at pace, creating a dazzling array of new products, services and capabilities. Each new generation of technology tends to offer not only greater speed, capacity and versatility in the transporting and processing of information–but also lower unit costs. The driving force of this revolution, of course, is the covergence of communications and computer technology which is literally redefining the telecommunications industry . . . . . Fundamental changes are starting to occur in our concepts of telecommunications. No longer is a telephone simply an instrument for voice communications but rather an information terminal, a means of assessing and exchanging information in a variety of ways, to and from a variety of sources. No longer is a television simply a passive instrument for receiving news and entertainment but rather an interactive communications device, visually displaying transactions that are taking place electronically. However, mentioning some of the complex issues that we face while we march through the development process, Brown said :

> How can we see to it that the information revolution does not bypass the lesser developed countries of the world, as was the case with the industrial revolution?
> How can we effect the transfer of communications technology from the haves to the have-nots under terms that are affordable to those who need it and equitable to those who invested so heavily to create it?
> How can we accommodate increased diversity in services and supplies without excessive fragmentation of national and international communications infrastructures?

In this context Kishore (1994) points out : the middle class elite in India whose life styles resemble that of the super rich anywhere in the world are at a comparative advantage over access to information and entertainment. The internal contradictions in present day India are stark clear; industrialised cities and marginally subsistent villages; jet planes and bullock carts; satellite technology and lack of clean drinking water, cellular telephones and absence of even one telephone in every village. This dualism has widened the gap between the rich and the poor.

An important study by Srivastava (1987) which has been frequently cited, to underline the socio-economic and political implications and issues, and future trends of television ownership and viewership in India in the year 2000, indicates that the rapid television expansion in the country will accentuate the knowledge gap between the poor in urban areas and the rich on the one hand, and between rural and urban people on the other. The study also suggests several measures to meet the situation including governmental intervention in the existing market structure so as to reduce the price of second hand black and white television sets for the poorer sections of the population, provision of community TV sets in rural areas and urban slums and introduction of TV sets in schools at public cost.

Even in the field of education, where with the introduction of new communication technologies, a teacher has to adapt to the role of a manager, facilitator and organizer of knowledge, there is an under current of scepticism among some educators that adoption of technology may lead to further elitism of education widening the gap between haves and have-nots. Adoption of technology needs lot of resources which every institution cannot afford. The government cannot provide every student with necessary facilities. This, it is feared, will create a new elite class rapidly, if technology is used selectively.

At the same time the perception at the policy making level and in the corporate world is that India should join the information super highway.

Delivering a lecture on Information Revolution through Satellite Communication, the former ISRO Chairman, Prof. U. R. Rao, said :

> The age of information super highway ushered in by the spectacular developments in satellite as well as terrestrial communication technology has, for the first time in the history of man, opened up a vast spectrum of opportunities which can transform entire societies. Given its talent, entrepreneurship and technical ability and through implementation of liberal national and proactive policies, India can surely derive the full benefits of information revolution. With the emergence of wide band multi-media communication between TV and telecommunication, India needed to formulate and adopt flexible, market place oriented regulatory measures through a

coherent policy based on the technological realities of the day. The challenges posed by rapid technological obsolescence can only be overcome through rapid communication, instantaneous access to information and establishment of adequate data base without which neither the technological establishment nor the industrial undertakings can compete in the globally integrated economy. A properly developed information networking system has become essential to all industries for not only accessing latest technological information but also for efficient management including optimal management of stock inventory, monitoring of technological progress and interaction with their own subsidiaries and other business partners in a shared video-conference mode.

Information on demand, video on demand, band width on demand and home television have become the buzz words of the age of information superhighway, implementation of which on a selective basis is essential to enable India to successfully compete in the global market in the liberalised atmosphere. With the liberalisation of telecom services through the enunciation of the new National Telecom Policy in 1994, many data services, audio and video-text services and video conferencing have now been privatised.

Privatisation, if properly implemented, can no doubt provide competent, highly reliable and user-friendly efficient service to the emerging giant India telecom market which is estimated to exceed 300 billion dollars in the next ten years.

REFERENCES

Agrawal, Binod C. and Arbind, K. Sinha (eds.) 1986. *SITE to INSAT: Challenges of Production and Research for Women and Children.* Concept Publishing Company, New Delhi.

Brown, Charles L. (1987). Revolution in Communication Technology. *Media Asia* 13 (1).

India, A Reference Annual (1996). Publications Division, New Delhi.

Kishore, Krishna (1994). The Advent of STAR TV in India: Emerging Policy Issues. *Media Asia* 20 (2) : 96-103.

Martin, William J. (1995). *The Global Information Society,* Aslib Gower, Hampshire.

Planning Commission (1997). An Approach Paper, Ninth Five Year Plan, Planning Commission, New Delhi.

Rao, U. R. (1996). *Space Technology for Sustainable Development,* Tata McGraw-Hill Publishing Company Limited, New Delhi.

Singhal, Arvind and Everett, M. Rogers (1989). *India's Information Revolution*, Sage Publications, New Delhi.

Srivastava, U. K. (1987). National Cost of Television in India in the year 2000. Socio-economic and political implications and issues. *Media Asia* 14 (3) : 136-148.

Voices (1997). Vol. I, No. 3-Special issue on *Telephone for All, Privilege or Basic Right?* Bangalore.

3

NEW INFORMATION AND COMMUNICATION TECHNOLOGIES AND SOCIAL IMPERATIVES

Since early 1960's Information Revolution has been the buzz word. The world has been experiencing, at an accelerating pace, an unprecedented Third Wave–fast forward-movement. The Information Revolution is characterised by information becoming the central and most significant commodity. An interesting off shot of the Information Revolution is–Informatization of Society on the one hand and the Commodification of Information on the other.

The significance of dawn of Information Revolution is underlined by the number of information-related terms it has given rise to : 'Information Society, Information Technology, Informatics, Information Industries, Informatization, Informatization Indicators and several more. Information has become a catalyst for economic development, and a key resource for literacy, education and social development. During the past five years or more there has been a flood of writings about super information explosion.' 'multi-media revolution' and 'information super highway', in relation to developments in Information Technology.

Information Society

The concept of Information Society goes back to the Japanese who are generally given credit for coining the Japanese term *Johoka Shakai* in mid-1960s. The term simply means Information Society, in which the primary activity of a majority of work force is in the areas of information production, processing and distribution. In the Information Society, information constitutes the basic resource and information workers are the main segment of employees. Whereas the social scientists consider Information Society as a process of technical change, subsequently it has been defined in terms of work force structure. In the Information Society, a majority of work force is engaged in the production, processing, transmission and distribution of information goods and services. The bewildering advances and developments in computerised information systems have brought the industrialised nations to the brink of what some have begun to describe as post-industrial era of the Information Society. The centrality of information/knowledge has also oeen emphasised by Daniel Bell.

In the post-industrial society, processing of information will be more important than production of goods. The value of information will be increasingly recognised, as well as the need for information specialists to act as guides to the vast information resources that will exist in the electronic form. The magnitude and diversity of the electronic resources will place trained information workers in great demand. The ascendancy of information as commodity is becoming apparent even in developing countries, in that majority of labour force in quite a few of these consists of information workers. Informatization is becoming a global trend. Information is considered not only as commodity but also as a cultural and social resource. The contemporary society is at a point where information and wealth are practically one and the same thing; information is commodity rather than a public service.

In his paper, Japan's Informatization Policy and the Urban-Rural Gap, Yutaka Dishi, *Media Asia* 16 (3), succinctly explains Information as under :

Information refers to the realization of an information society, which in turn, is defined as follows :

An Information Society is a society characterized by abundant information in terms of both stock and flow, quick and efficient distribution and transformation of information, and easy and inexpensive access to information for all members of society, (Ito, 1980). The Ministry of Posts and Telecommunications (1985) also indicated some of the key features of informatization as follows :

(a) Public needs for information have become increasingly sophisticated and diversified,
(b) New communication media share certain border areas with conventional media such as mail, telecommunications and broadcasting, and

(c) Networks (i.e. data communications) are being formed on a large scale.

The National Information Technology Plan for Singapore defines Information Technology as the use of computers, telecommunications and office systems in all aspects of information flow in Singapore from collection and processing to storage, package and dissemination. Information Technology, therefore, constitutes an integral component of the economic development plan of Singapore. It plays a dual role in the economic development of Singapore. It is a high tech industry in itself creating IT related products and services, as also a tool to increase productivity and enhance competitiveness in all sectors of society.

Alvin Toffler and Third Wave Media

The Information Age *guru* Alvin Toffler considers contemporary societal changes and technological developments as high speed revolution, the Third Wave, which is creating a new civilization all around us. Classifying agricultural phase as the First Wave, industrial phase as the Second Wave, he considers exploding change–with personal lives being torn apart, the existing social order crumbling, and a fantastic new way of life emerging on the horizon, as the Third Wave, with new life styles, new values and technologies, new modes of communication, new ideas and concepts emerging all around.

The changes that Toffler envisions are indeed sweeping. Taking up the analogy of the Third Wave, in a fascinating account, Toffler has classified de-massified media as the Third Wave Media.

The burgeoning flock of mini-circulation weeklies and bi-weeklies that serve the neighbourhood communities as against mass circulation dailies, leading to segmentation of the reading public, larger number of radio stations leading to specialised audience segments and citizens band radio making it possible for the listeners to communicate, Cable Television which de-massifies the audience–all constitute the Third Wave Media. The Video Cassette players and recorders; and domestic satellites making it possible for individual television stations to form mini-networks–leading to smaller audiences and a larger multiplicity of programmes are the other constituents of the Third Wave Media—giving rise to new info-sphere alongside the new techno-sphere. The Third Wave media analysis of Toffler underlines the diversity and intensity of

Information Society wherein entertainment, information and leisure are dominated by the products of Information Technology.

New Information/Communication Technologies

It is fascinating to observe that whereas the developing countries are still coming to terms with socio-economic and cultural implications of proliferation of mass media broadly classified as print and electronic media; the rapid advances made in Information Technology products, (the hardware which is the backbone of information society), New Media emerged in the late 1970. In the paper, *Information Society in their Mind*–A survey of college students in seven nations *(Asian Journal of Communication,* I (II), 1991) Georgette Wang lists the ten items as New Media : one way and interactive cable television, video cassette recorders, video disc players, videotex, teletext, facsimile, personal computers, direct broadcast satellite and cellular radio.

The study, essentially explorative in nature revealed that awareness and knowledge among college students of different countries about New Media and Information Society varied greatly and reflected the extent and the degree to which the economies of their countries were industrialised. The variation is reflected in the observation : It is no secret today that even industrialization is still a long way from reality for a great number of nations in the world, not to mention informatization. To these nations, informatization will not likely become a focus of attention before development has reached a level at which the production of information and information machines plays a key role in national growth. When viewed against the current world development trends, it is distinct possibility that the information age may be a long, long way off for many societies.

Ten years after the opinion survey was completed, the situation has substantially changed. The New Media, particularly in India–facsimile, personal computers, cellular telephones–have expanded fast. Their application in various fields-teaching and training, literacy, distance education, financial institutions and other spheres has become increasingly conspicuous.

The New Media, referred to earlier, are now generally classified as New Information and Communication Technologies (NICT). In his paper, New Information/Communication Technologies : Their Role in Education (*Media Asia,* 22 (2), Kiran Karnik lists photocopying, video recording, facsimile, personal computers, E-mail and computer networks, satellite, telecommunication, including mobile communication and satellite broadcasting and receiving equipment for satellite transmissions as NICTs.

Some of these, like photocopying have been in use for many years, what is new is their present form, speed, accessibility and user-friendliness. However, with the phenomenal advances in communication technologies, the concept of 'new' is becoming controversial. This is particularly so because changes are so spectacular that what is new today may become archaic tomorrow, owing to the rapid changes in the industry. It is also apparent that more and more writers on communication technology are using the term 'Information and Communication Technology' or 'New Information/Communication Technologies', inter changeably, rather than separate the technologies of information and communication. The convergence of both the technologies is such that it is difficult to distinguish the respective product application of one from the other. Each of them serves various functions but they share some common features. They disseminate messages with greater speed, accuracy and efficiency.

Multimedia

Today, it looks as if the word multimedia is sweeping the globe. Technologists, planners and policy makers seem fascinated to talk about the bright future ahead–the dream world of the 21st century with the utilization of multimedia. Multimedia refers to the linkage of computers, television sets and telephones in an increasingly elaborate global network. The electronic world is referred to as information highway, because of its potential to tap a vast realm of digitized information : stock market information, magazines, shopping catalogues, libraries– much of which is available on the Internet. It is the boundless worldwide web of computer networks.

The most commonly stated advantage of multimedia for Asia and the Pacific countries is that because it enables any remote region to receive information without delay, its utilization will be most useful in countries with vast land areas or mountainous and complex topography. Moreover, in the multimedia society receivers can also act as senders, whereas traditional media, such as TV, radio or books can only carry information one way, from senders to its receivers. Such an unprecedented Information Revolution will change the flow of people, their awareness and even social structures.

It should, however, be noted that it takes a tremendous amount of money to build the infrastructure necessary for the application of multimedia and equally huge amounts of energy to maintain the systems.

Information Technology

Today's advanced information communication technologies are not only the key carriers and processors of information content, but are increasingly vital technologies for economic development. They provide an indispensable infrastructure for all industrial production processes. They are also 'synergetic' i.e. they lead to growth in many other industries. They create an elaborate infrastructure around their products and services, similar to the automobile industry earlier in this century.

The US Vice-President Gore advocates the "National Super Highway", to be realised by the year 2015. Likewise in Japan there is a plan to spread the same type of network system as US by 2010. Similar plans have also been announced in Europe, China and Russia.

NICT and Social Imperatives

It would be interesting to recall the glowing euphoria, that accompanied the launching of the Early Bird Satellite by the US scientists in May 1965–heralded by the *Time* newsmagazine in its cover page story (May 14, 1965)–The Communications Explosion. Visualising a startling communication explosion ahead, the science feature on Electronics succinctly predicted a series of developments in medicine, world-wide telephoning, data facsimile transmission, educational TV and several kinds of entertainment offering choice of programmes and more. At the same time the glowing euphoria was tempered with caution. It was pointed out; whether or not the spread of such scientific largesse will indeed "save the world" is a problem that would not be solved by scientists alone. The sociological implications are immense. The analysis added : "I doubt if more food will be grown in India," says RAND Corp. Sociologist Joseph Goldsen, "even if every village gets a television set with lecturers teaching new agricultural techniques every hour. It takes generations to change customs and traditions."

This precisely is the dilemma that modern day communication scholars, sociologists, planners and policy makers have to face— whether, and the extent to which information and communication technologies could serve the Third World countries in confronting their social and economic problems. If any, the scepticism and dilemma is being articulated with increasing concern.

New Communication Technologies and African Culture– Preservation or Pulverization

In a detailed presentation as to whether NICTs would help preserve the African cultures or assist in their pulverization, Jerry Komia Domatob says : "With the advent of new communication technologies–computers, cables, satellites and video cassettes two diametrically opposed views are advanced. While pessimists and cynics contend that they may accelerate Africa's cultural pulverization, optimists contend that they could be used to preserve African cultures. Both positions reflect some truths about new communication technologies and the preservation of African cultures. They may actually erode African culture and introduce new values. Since the new technologies are packaged with their own value systems–language and work ethic-it can be submitted that they may annihilate facets of African culture. In fact, they can lead to cultural synchronization."

There is a proliferation of personal computers and audio and video cassettes in the continent and other affordable technologies are being acquired at a phenomenal rate. In addition signals from Europe invade television stations in all parts of Africa, and European programs now inundate the Maghred states of Egypt, Libya, Tunisia and Morocco. More than anything else, multinational corporations such as IBM, Thompson CSF, RCA, Sony and Panasonic transfer these new technologies to Africa with remarkable speed. Although the walkman is a new technology, it now dominates African markets. Africans cannot, therefore, evade this new trend. It is verily a phenomenon that has come to stay, be it good or bad.

Cultural Synchronization

The fear of cultural synchronization, in the wake of globalization of trade in cultural products and increasing domination of USA in the production, distribution and consumption of cultural and media goods has been frequently articulated. Cultural synchronization essentially implies that the traffic of cultural products goes massively in one direction, consequently local inventiveness and creativity of the recepient culture is thrown into conflusion or destroyed. This fear has been further accentuated when we take note of the surge in the overall American trade position of audio-visual products. After all the cultural industries, which include movies, TV serials are 'second' only to the aerospace industry in terms of their contribution to the American exports. To quote *Media Development* (1994, p. 4) :

> The scramble for the media and information market is reminiscent of the scramble for territory in colonial times. Except that this war is being won and lost in corporate board-rooms, on the stock market, in the financial corridors of the world and finally in the hyper-saturated media environments of our lives.

The Information Technology products have greatly contributed to this fear amongst Third World countries, which frequently finds expressions in the writings of media scholars and policy planners.

Technological Leap Frogging–Satellite Technology

Eversince the launching of the Early Bird Satellite in 1965 the satellite technology has been the prime mover in accelerating developments in the area of communication technologies. The Third World countries find satellite technology particularly relevant to their development needs as it helps them leap frog obsolescent technologies.

Arthur C. Clarke, the eminent author of the science fiction book *Movie 2001: A Space Odyssey.* Distinguished scholar and science writer was the first to visualise the potential of satellites for the developing countries. Developing countries stand to gain most from these innovations as they are not yet burdened with heavy fixed investments in technology that is rapidly becoming obsolete. In one of the articles—New Communication Technologies and the Developing World, Clarke wrote :

> But why do we need such huge satellites, what have they got to do with the problems of the Third World? The answer may seem paradoxical, even perverse. In highly developed regions like the United States and much of Europe, communication satellites are a great convenience but are not absolutely vital. These countries already have excellent cable and microwave links.
>
> To many developing countries, however, satellites are essential, they will make it unnecessary to build the elaborate and expensive ground systems required in the past. Indeed, to such countries, satellites could be a matter of life and death. To put

it as dramatically as possible, unless major investments are made in space, millions are going to die, or eke out brief and miserable lives. And most of those millions will be in the Third World.

The eminent Indian space scientist and visionary Vikram Sarabhai, who was primarily responsible for leading India into a modern technological direction said : A developing nation following a step by step approach towards progress is landed with units of small size, which do not permit the economic development of new technologies. Through undertaking ventures of uneconomic size with obsolete technologies, the race with advanced nations is lost before it is started. Developing nations such as India have the possibility of effectively using space communications for national needs. Compared to advanced nations they have indeed an advantage though not having an existing major investment in older technologies.

One of the extensively documented and discussed instances of use of satellite technology for leap frogging several stages together has been through the use of Palapa Satellite in Indonesia. The significance of launching of Palapa Satellite system in 1976 has always been considered strategic because it was able to overcome Indonesia's vast distances and link its administrative units. The underlying consideration for the Indonesian government to introduce satellite communication was to support national development activities.

Palapa was considered as an experiment in rural development through use of advanced communication technology, commercial telephony, telex and broadcasting. It was reckoned as the most cost effective and quickest way of linking the 150 million people of Indonesia. The system was used to launch development communication programmes in areas like family planning, improved health and hygiene and increased agricultural production.

It is, however, interesting to observe that the claims of successful results in terms of learning, narrowing the knowledge gap between the lower and upper social and economic strata, promotion and adoption of family planning and modern health care, etc. are now being questioned. It is pointed out that it may be specious to attribute instances of economic progress made by individuals primarily to their owing a TV set alone. A study by Holaday undertaken as a part of research project (*Media Asia* 23 (2) : 1996) points out that the design of an earlier study by Chu *et al.*, did not go into the issues specifically as to whether claims like

Relevance of Communication Satellites to development : claims	***Reconsideration of these claims***
1. National integration Many Third World countries. either by reason of their immensity and geographical dispersal like India and Indonesia, or by reason of their heterogeneous social composition or because of natural barriers like vast expanses of forests, deserts or mountains are yet to be integrated into a single polity. Domestic distribution satellites can perform this function much more cost effectively.	1. National integration Lack of national integration is far more complex than is suggested by territorial vastness or the incidence of forests, mountains and deserts. Very often there is need for more autonomy for the constituent national entities. In such situations satellites can be used either as instruments of coercion or as a substitute for more human forms of achieving integration.
2. Administrative effectiveness As a corollary to 1 above, administra-tive integration and effectiveness is hampered. The bureaucracy tends towards concentration and higher effectiveness in the metropolis, and declines outwards.	2. Administrative effectiveness Bureaucratic concentration in the metropolis, which is largely respon-sible for administrative ineffect-iveness in the Third World, is likely to be aggravated through a comm-unication satellite.
3. Delivery of education, both formal and non-formal including teacher training The shortage of schools, teachers, equipment and buildings is chronic in most Third World countries. Satellites have the capacity to multiply these meagre resources at a fraction of their terrestrial costs.	3. Delivery of education This slogan has been used consistently to justify the introduction of TV into the Third World. TV is recommended to the Third World not for entertainment, or for commerce, but for education. For over two decades experiments have been conducted for determining the adequacy of TV as a vehicle for education. However, there is consi-derable evidence to show that education is best delivered in the face-to-face inter-active situation.
4. Delivery of agricultural extension As in the area of education, effective agricultural extension work is hampered by the lack of trained extension workers, the lack of transport, the inability to have as many demonstration plots as are needed, etc. All this can be overcome through satellites.	4. Delivery of agricultural extension The problem of agricultural extension in Third World countries is more than a question of access to knowledge, and demonstration. It is primarily a question of land tenure, the lack of credit, the lack of irrigation facilities, the high cost of inputs, etc. Most of these are structural and political questions.
5. Delivery of family planning programmes It is claimed by many development economists that the greatest single impediment to development in the Third World is population growth	5. Delivery of family planning programmes The consideration that apply to education apply here too.

which every year either outstrips or at least equals economic growth. Family planning is therefore, seen by them to constitute the center-piece of any long term economic strategy. But the crucial question underlying all family planning programmes is the one concerning communication. It is claimed that satellites are the answer.	
6. Delivery of medical and health care services What has been said of educational, agricultural and family planning extension is equally true of medical and health care. The same infra-structural and personnel inadequacies baulk the delivery of adequate diagnostic and curative services from the metropolis, where a certain volume of professional skills is concentrated, to the outlying provinces where they are grossly inadequate or totally absent.	6. Delivery of medical and health care service. The fundamental issues of medical and health care in the Third World are not related to the lack of communication but have to do with the lack of clean drinking water and the lack of protein and vitamins. They are not diagnostic but directly stemming from poverty.
7. Isolation from marketing information, both national and international A modern economy must have an access to supply data, the most up-to-date, instantaneously and on call both to suppliers and customers, abroad and at home. Unless entrepreneurs, banks and firms have access to data banks and facsimile services through computers linked to satellites using digital transmissions, the international market would collapse. On a much smaller scale, this would be equally true of the domestic market.	7. Isolation from marketing information This is a disability peculiar to big entrepreneurs and multinational corporations based in Third World metropolitan centres and not a felt need among small-holding subsistence farmers and village cooperatives who constitute the bulk of the rural sector. The solution to the latter's need for fair prices is not likely to be provided by instantaneous satellite communication.
8. Participation as a development tool Most Third World societies are highly hierarchical. Planning and decision making are centralised and removed from the people. There are neither political nor societal mechanisms that enable the widest possible community to participate in either the political or the production process. This is one of the reasons attributed to account for low productivity in thse societies. Normally, securing participation in broad societal processes is a highly complicated enterprise in social engineering. But technologists see in communication satellites the final answer.	8. Participation Like pluralism, participation is a product of political consciousness, of the pressure of organised and motivated people on the structures of power. It can never be contrived, least by recourse to technological devices.

The above table is reproduced from the paper, "Media Technologies and Socio-Economic Imperatives–An Indian Perspective"–Vir Bala and V. S. Gupta, Published in *Communication Planning for Socio-Economic Development* (1994).

viewers taking loans from public institutions as against non-viewers getting loans from private, high interest money lenders; viewers doing business in the regional markets, whereas non-viewers selling their goods through intermediaries, and more viewers joining cooperatives and development projects than non-viewers; only because of the introduction of television, were valid.

In a scintillating presentation on Communication Satellites–A Third World Perspective, made at a colloquium held in Bonn in 1982, a leading Third World Communication researcher, Neville D. Jayaweera, argued that assumption underlying the proposal to use communication satellites in the Third World were wrong and satellites could not provide solution to the problems which were primarily political, economic or sociological. A brief summary of his views on, Relevance of Satellites to the Development-Communication Debate, is given in the tabular form. In his presentation he expresses his apprehensions that the satellite technology after all may not have the potential that it is credited with; on the other hand it might consolidate and perpetuate the inequities rather than evolving a redressal mechanism.

The views of the two contending schools of thought, those who consider satellite technology as a panacea for solving major problems of the country, and others who consider development more than merely a technology-driven, complex process, have been given in detail only to underline the dilemma, and far reaching implications of new information and communication technologies.

Cable Television

Much like the above satellite technology, development of cable television was also heralded as 'an answer to the need for more diversified content', since cable television offers the possibility of unlimited number of channels for a multiplicity of users and purposes (multiple channels, multiple choices for the viewer) it was trumpted as 'technology of abundance'. It was, however, not adequately realised that cable television could not be optimally used unless there are a number of stations with a variety of programme material available to feed those channels. In the absence of sufficient quantity of culturally compatible, civic and community oriented software available, the cable operators have to fall back upon whatever is available in the entertainment supermarket.

Let us not delude ourselves that the problem of cultural alienation through cable television is basically limited to the elite class; because it is the elites who guide and govern the important decisions in the government and it is important to consider how their value system is developed.

In the Report of the International Commission for the Study of Communication Problems, it has been stated that introduction of cable television has profound effects on the existing broadcasting situation. Capital investment in cable hardware may constitute "more of a privilege of the affluent urban dweller than a realistic option for more than three quarters of the world's population". Some countries that have introduced cable television note its detrimental effect on "cultural sovereignty."

Some of the observations in the Report are so interesting that these need to be reproduced in detail :

The promise of cable television (multiple channels, multiple choices for the viewer) may well have been oversold in its early development period. Although the technology lends itself to a wide choice of programmes that can serve individual and minority needs, it also tends to splinter the national audience into many mini-audiences, which is not in all circumstances viewed as beneficial. Moreover, the cost of providing separate programme material for three dozen or more channels all at once is beyond the means of many such enterprises.

One of the most favoured aspects of cable television is the availability of channels for educational and community access programming. In some countries where cable has been introduced, the regulatory authorities have made it a prerequisite that cable system operators, whether public or private provide a community or public access channel which citizens of the community can use for local interest programming. There were high expectations that community access channels would enrich community life by fostering communica-tion among individuals and community groups; provide a new forum for neighbourhood dialogue and artistic expression, revolutionize the communication patterns of service organisations, consumer groups and political parties. In effect, cable television's provision of the community access channel was hailed as the technology that would personalise the electronic media and create a sense of community in urban settings, which had been depersonalised by the advances of other technologies. (p. 89-90)

The Information Society–A Retrospective View

In their seminal book, the *Information Society–A Retrospective View*

(1993), Herbert S. Dordick and Georgette Wang have made an attempt to revisit the claims that have been made for the emerging information society. One of the paragraphs reproduced below heralds the dawn of a new milennium, a brave young world, full of promise and prosperity.

The explosive diffusion of computers, telecommunication, broadcast and video media, and the information media is seen in national and global statistics. Almost half of the workers in the industrialized world are employed by the rapidly growing information industries, and in almost every industry there are workers, whose primary tasks are to create, collect, process, and distribute information. The value added information products and services to the United State GNP is today slightly greater than 45 per cent and for Europe and Japan somewhat more than 40 per cent. The world information industry, even if narrowly defined, will be greater than $ 500 billion before the end of the decade. The production, processing, and distribution of information is a major industry throughout the world. By the year 2000, it will account for about 40 per cent of the world's industrial production.

However, doubts and scepticism sets in almost immediately as to the efficacy of the newly emerging dispensation to the Third World nations. The nagging issues of widening gulf between the information-rich and the information-poor as also between the haves and have-nots start emerging as under :

If there is a growing gap between the production and consumption of information there is, perhaps, an even greater gap between what have come to be known as the information-rich and the information-poor. Information technology and modern telecommunications raised renewed hopes for enhancing national development. Development information replaced development communications as the mechanism for stimulating growth in the developing and newly industrializing countries. Whether development information can be a catalyst for development is a moot point in many countries, where reliable telephone or even telegraph communications are rarely available, and where computers are only in the dreams of young people returning from studying in advanced countries. Plagued by barriers of poverty, poor health, low life expectancy, illiteracy, and little access to tertiary education, emerging countries must compete with advanced countries in a highly competitive global marketplace. What does the information society mean to them? Economic growth in the least developed nations of the world has stagnated or has been increasing very very slowly.

The gap between the developed and lesser developed nations continues to widen annually. Can these Third World nations ever catch up.

Technological Dilemma

The planners and policy-makers throughout the Third World are faced with a dilemma, as to the extent to which new information and communication technologies can be usefully harnessed to solve the problems of under development. To resolve this dilemma, bold decisions and options will have to be taken, the choice to be made are economic and technological but also political in the first instance. Political decisions must be made in order to avoid some of the unforeseen economic and technological implications.

Two members of the International Commission for the Study of Communication Problems, Gabriel Garcia Marquez and Juan Somavia in their general comments to the Report have said : There is a tendency in different parts of the Report to "glorify" technological solutions to contemporary communication problems. We want to emphasize that the "technological promise" is neither neutral nor value free. Decisions in this field have enormous political and social implications. Each society has to develop the necessary instruments to make an evaluation of alternative choice and their impact. The insistence on the need to develop communication infrastructures in Third World countries is correct and necessary, but it should not be overstated. It is not possible to solve contemporary communication problems through money and training alone.

REFERENCES

Atal, Yogesh (1987). The Communication Revolution in Asia. *Media Asia* 14 : 3-6.

Bhattacharje, Ajit (1995). National Media Policy–New Technology and the Press. Paper presented at the AMIC-India FES-IIMC Seminar, New Delhi, November, 1995.

Chan, Joseph Man (1991). Communication Networks in the Asia-Pacific : An Overview. *Media Asia* 18 : 35-51.

Clarke, Arthur C. (1981). New Communication Technologies and the Developing World. *Media Asia* 185-190. (as reproduced in *Opening Windows-Issues in Communication,* The Silver Jubilee Collection, AMIC Singapore).

Dhanarajan, G. *et.al* (1990). Findings–Staff development for the application of interactive technologies in distance education–A feasibility study. *Media Asia* 17 : 203-211

Doatob, Jerry Koia (1990-91). New Communication Technologies and African Cultures: Preservation or Pulverization. *Caejac Journal* Vol. III, 60-73.

Dordick, Herbert S. (1987). The Information Technology—'Revolution' and National Sovereignty. *Media Asia* pp. 123-129.

Dordick, Herbert S. and Wang, Georgette (1993). *The Information Society–A Retrospective View.* Sage Publications, New Delhi.

French David and Richards Michael (1996). Communication and Human Values Series—*Contemporary Television Eastern Perspectives.* Sage Publications, New Delhi.

Fukuyama, Francis (1995). *Trust : The Social Virtues and the Creation of Prosperity.* Hamish Hamilton, London.

Goonasekera, Anura (1993). Media Technology and Social Imperatives : An examination of Communication Policies in Selected Asian Countries. *Media Asia* 20: 194-204.

Ishadi, S. K. (1989). Towards the Information Revolution Era—Using Satellites in Indonesia. *Media Asia* 16: 41-46.

Jamias, Juan F. (1993). The Impact of New Communication Technologies on Cultural Identity in Rural Asia. *Media Asia* 20 : 205-207.

Jussawalla, Meheroo (1993). Information Technology and Economic Development in the Asia Pacific. *Media Asia* 20 : 127-132.

Mowlana, Hamid and Wilson, Laurie J. (1988). Communication Technology and Development. UNESCO 101.

Pereira, Oswald (1996). Connectivity Links Unlimited, *Outlook,* July 10, 1996, pp. 42-43.

Rahim, S. N. (1987). The Nature of the Telecommunications Revolution in the Developing Countries of Asia–Its Impact on the Economy. *Media Asia* 14 : 39-45.

Rao, Sandhya and Natesan, N. Chinna (1996). Internet : Threat or Opportunity for India, *Media Asia* 23 : 96-99.

Samar Halarnkar (1996). Cowboys of Cyberspace. *India Today,* June 30, 1996, pp. 131-33.

Samarajiva, Rohan and Shields Peter (1993). Integration, Telecommunication and Development : Power in the Paradigms. *Asian Communication Handbook,* Anura Goonasekera and Duncan Holaday (eds). pp. 359-380.

Sinha, Nikhil (1995). Telecommunications, Capabilities and Development : Towards an Integrated Frame Work for Development Communications. *Media Asia* 22 : 17-25.

Time (1965). The Communications Explosion : Early Bird and After: Electronic—The Room Size World, pp. 48-51, 14 May.

Time International (1995). Can Japan Catch the Wave? 27 February. pp. 34-39.

Toffler, Alvin (1981). *The Third Wave.* Pan Books Limited, Cavaye Place, London.

UNESCO, (1982). *Many Voices, One World.* Communication and Society–Today and Tomorrow.

Wang, Georgette (1991). Information Society in their Mind : A Survey of College studetns in Seven Nations. *Asian Journal of Communication* I :1-18.

William Shawcross (1994). Murdoch–A Review Appearing in the *Hindu,* 4 September, *The Media Predator.*

4

NATIONAL MEDIA POLICY

The International Commission for the Study of Communication Problems in its Report (1982) recommended : Communication be no longer regarded as an incidental service and its development left to chance. Recognition of its potential warrants the formulation by all nations, and particularly developing countries, of comprehensive communication policies linked to overall social, cultural, economic and political goals. Such policies should be based on inter-ministerial and inter-disciplinary consultations with broad public participation.

Stressing the importance of Communication policies, the Report said : Communication policies and development strategies, considered as essential means of solving the major problems of our time, should be designed first and foremost to ensure that the media of "information" become the media of "communication". Because communication presupposes access, participation and exchange, different media should be involved in the progress of democratizing it is here that the links between communication, the democratization of communication and the democratization of development become apparent. For if communication policies are to be aimed at democratizing communication, and if, in addition, they must essentially be designed to be integrated into development programmes, then it is vitai for those concerned to participate in the work of programming and decision-taking. There are many ways of achieving this, provided that the political will exists in the first place.

The debate on formulating a National Media Policy has been a constant preoccupation with the Ministry of Information and Broadcasting, media academics, professionals, and non-governmental organisations.

The Government of India, over the past more than 25 years, has formed several committees to study various aspects of media policy : autonomy of electronic media, software production, distribution of air time, cable television, etc. The Consultative Committee attached to the Ministry of Information and Broadcasting in its meetings held in July 1993 also had National Media Policy as one of the agenda items for discussion. The Committee was of the view that a National Media Policy be forumulated; although there were differing opinions regarding its form and contents. Subsequently, the Ministry circulated a Note on National Media Policy–A Discussion Paper, and a Questionnaire to elicit opinion from a cross section of society, including media professionals, experts and the general public.

In the Foreword to a book *Public Policy, Analysis and Design,* the eminent Indian diplomat, economist and educational administrator, Prof. A. M. Khusro has said, "Policy need not be made by governments only, but in this country policy has been made almost wholly by governments. It may be because other segments, sectors and socio-economic agents in the country–the private industrial sector, the commercial sector and the social sector, have been weak and are in the throes of a slow developmental process, they will acquire a degree of resilience and a policy making capability later. But to begin with they were weak and therefore, it fell to the lot of central and state governments to make policies and programmes during the last 40 years. . . I have already stated that Non-Governmental Organisations (NGOs) that could make policies are still weak but that they are emerging slowly. There are NGOs like Policy Initiatives or Policy Alternatives, etc. and there are the associations of industry and trade. Many people in the private sector of industry and trade, the cooperative sector, the NGO sector, the social sector and the intellectual sector, are beginning to think of many policy areas and it is certain that their role will increase with time.

At the very outset, it may, however, be elucidated, that in literature pertaining to Media Policy several terms have been used : Communication, Communications, Media Policy/Policies, and Communication Policy/Policies.

In this context, the International Commission, referred to above clarified : Definitions of several general notions dealt within this Report–more particularly, those of communication and information–have not been standardized nor is there a unanimity as to their meanings. In fact, the terms are used by different authors in a variety of ways, with many nuances regarding their components and significance.

The Discussion Paper on National Media Policy prepared by the Ministry of Information and Broadcasting also stated that the present policy has been variously termed by both the government and those demanding it at various points of time as a Media Policy, an Information Policy and as a Communication Policy. Needless to say, all these convey possibly different meanings.

In the absence of any standardization, an effort has been made to retain these terms in the context of the relevant texts.

The Chairman of the Press Council of India, A. N. Sen has, however, been of the opinion that it may not be wise to seek to lay down any firm National Media Policy, as the effective implementation of the Policy may create more problems than it is likely to solve; although there may not be any difficulty in arriving at a general consensus as to certain norms to be observed by the media instead of seeking to lay down any firm National Media Policy. At the same time, however, Sen observed :

> A policy is generally formulated and adopted for achieving a particular object. In the absence of any such object or fixed goal sought to be attained, there can be no question of framing any policy. A policy adopted for achieving no particular purpose will indeed be meaningless. To properly understand and appreciate whether any National Media Policy is at all necessary at this juncture, we have to bear in mind two important considerations : (1) whether the same can be adopted without interfering with the independence of the media; and (2) exactly what is the object which the media are expected to achieve and what is the purpose which the media seek to serve.

Objectives of National Media Policy

In a country as vast and diverse as India, with a number of languages and still greater number of dialects, regions, varying levels of socio-economic development of its population, the media would essentially have pluralistic character in full measure. How should then the media serve the essential objective of imparting information, education and entertainment, apart from meeting the requirement of national development. A number of academics, expert committees and working groups, have from time to time tried to lay down broad guidelines as to the objectives of National Media Policy. Some of these are :

1. The media structure of each country is dictated by its own special needs and national objectives. Whereas all developing countries have, more or less, similar and common media requirements and objectives, each country may have its special problems and needs. Referring to these, D. R. Mankekar points out : India is a country of sub-continental size, with a federal structure and gigantic population, with a multiplicity of distinctive cultures, languages, races and religions, with three-fourths of the population living in the villages. It is specially concerned with problems of national integration, political stability, population explosion and liquidation of poverty and illiteracy. These problems must inevitably receive the highest priority in India's programme of nation building, and therefore, the fashioning of a communication media infrastructure, for India must have for its objective a concerted attack on these problems. Most of these problems are also common to all the developing countries of Asia, though in varying emphasis and degree.
2. National Media Policy, according to Pran Chopra, should help media fulfil the objectives of providing access to information to the largest number of people, preserve their credibility and work as two-way channel. He defines National Media Policy as under : By national policy I am only going to mean those decisions of the national government which might help the media fulfil functions that are appropriate for them in a democratic and pluralistic country like India. I consider four functions of the media to be the most appropriate in the Indian context. First, the media should provide to the largest number of people easy access to information which is reliable, relevant, and abundant. Second, the media must provide reasonably free channels of expression for opinions which are well informed, responsible and adequately reflect all relevant points of view. Third, media should be a two-way channel : not only addressing messages to the audience but also giving the audience a forum for the expression of its concerns. The fourth overrides all these three functions : the media must preserve their credibility.
3. In a paper, National Communication Policies, media educator M. R. Dua says : Generally speaking, the objectives of national communication policies are to inform masses about the developments taking place in national and international spheres

affecting their daily life. These policies also help inspire, motivate and enthuse the people to support and cooperate in nation-building, economic progress, social development, imbibe national cultural values and education. Communication policies also assist coordination and cooperation among multifarious national and international agencies; promote rational thinking and deal with national problems in a logical, systematic manner. The most crucial objective, however, is to bring about overall wider acceptance of programmes for national development keeping the fast changing global communication scenario.

4. The Sub-committee of the Consultative Committee of the Ministry of Information and Broadcasting spelt out the objectives of National Media Policy in its Report, submitted on 29 March 1996. According to the report : The National Medial Policy should lay the broad framework of the relationship of the Government with both national and international media such as Television (domestic, foreign and cable), Video, Internet, Radio, Newspaper, Film, etc. Its broad objective should be to help a citizen of the country to realise his or her potential best. The reach of print media is limited to the literate people only, whereas the Electronic Media–Radio and Television–has entered the households of the people all over the country. Thus, it has the viewership of all kinds–adult as well as minor, illiterate as well as literate, poor as well as rich. Slightly different is the approach of film media which has also the viewership of both illiterate and literate, but its viewership is basically voluntary. Its visual impact though is much more than that of print media on both educated as well as uneducated.

Accordingly, the Sub-committee proposes the following objectives :

- Not to permit growth of monopoly in any media.
- To project the developmental needs and social, cultural and economic aspirations of the people, particularly in the rural areas who constitute a bulk of our population.
- To strengthen, preserve and promote our democratic traditions, culture and values, national integration and scientific temper.
- To promote and strengthen sense of national integrity built on secularism, socio-cultural pluralism and linguistic diversities (unity in diversity).

- To facilitate greater access to information, education and entertainment in that order to all sections of the people,
- To enable development in the wake of challenges thrown up by rapid technology.
- To provide wider choice to the people in matters of information and entertainment.
- To act as catalyst for social change.
- To broadly ensure that discrimination based on gender, religion, caste, language, etc. is not directly or indirectly practised or promoted and portrayal of any section of the society is not done in a derogatory manner.

The National Media Policy, in the Indian context, is governed by the provisions of the constitution regulatory and legislative measures, policy guidelines for official media and the codes evolved by the professional bodies. These guidelines, alongwith the Press Council Act, the Cable Television Networks (Regulations) Act, 1995, Supreme Court Judgement on Freedom of Airwaves (February 1995) constitute the core of National Media Policy in the country.

An overview of National Media Policy is greatly facilitated by the deliberations, reports and recommendations of several committes and working groups constituted by the Government of India to study in-depth various issues impinging upon communication, development, regulatory measures, etc.

ELECTRONIC MEDIA POLICY

1. Committee on Broadcasting and Information Media (1964)

This Committee, also popularly known as the Chanda Committee, constituted in December 1964, presented the Report on Radio and Television in April 1966. The Committee observed : It is not possible in the Indian context for a creative medium like broadcasting to flourish under a regiment *(sic)* of departmental rules and regulations. It is only by an institutional change that AIR can be liberated from the present rigid financial and administrative procedures of the Government. It recommended that the separate corporations, set up for Akashvani and Doordarshan, should have the freedom to evolve its own methods of recruitment, regulate scales of pay and conditions of service according to its needs, and devise a financial and accounting system appropriate to its creative activity.

This report was considered by the Cabinet in December 1969,

and the Lok Sabha was informed in April 1970 that the present was not an opportune time to consider the conversion of All India Radio into an autonomous Corporation. However, another important recommendation that Akashvani and Doordarshan be separated, was accepted and took effect from 1 April 1976.

Arguing for the grant of autonomy to the two media, the Chanda Committee had said : Theoretically it (AIR) has the freedom of a national newspaper to present objectively topics of current interest but in practice it has failed to do so mainly for two reasons. First, successive ministers usurped interfering even in matters of programme planning and presentation, and second, the selection of directing staff was so made as to ensure unquestioning compliance. As a result the public image of AIR has become tarnished, its objectivity destroyed and its initiative gradually whittled away.

2. Working Group on Autonomy for Akashvani and Doordarshan (1977)

The Committee, also popularly known as Verghese Committee, appointed in August 1977 submitted its Report to the Government in February 1978. The Report is a landmark document on the aspect of granting autonomy to the electronic media. Its major recommendations were :

(i) An autonomous national trust should be established under which Akashvani and Doordarshan would function. It gave a name to the proposed authority–Akash Bharati : The National Broadcast Trust,

(ii) Radio and TV should work for the public purpose. They should function within the framework of a broad perspective of national communication policy,

(iii) The proposed autonomous broadcasting trust authority should be owned by the nation and be accountable to the Parliament,

(iv) The priority of broadcasting has to change from 'urban-elitist moorings to the rural and semi-urban areas and to the urban poor'. The quality of programmes would have to improve. It should also work towards filling the rural urban and tradition modernity gaps,

(v) The idea of one autonomous corporation each for All-India Radio and Doordarshan did not find support from the Working Group. It opted for one autonomous national trust for both of the broadcast media—radio and television. It also rejected the concept of autonomous regional corporations, but it envisaged decentralization of national broadcasting authority, and

(vi) The autonomy of the authority and independence from control of the government should be guaranteed by the Constitution.

The Government introduced a Bill in May 1979 which envisaged setting up of an autonomous corporation known as Prasar Bharati. Since the Government went out of office, the Bill lapsed and the successor government headed by Mrs. Indira Gandhi decided not to set up an autonomous body.

The Government later on introduced the Prasar Bharati (Broadcasting Corporation of India), Act in September 1990. However, after the prolonged foot-dragging on the implementation of the Prasar Bharati Act, the Bill was notified only on 22 July 1997, seven years after it was enacted. It came into force from 15 October 1997; subsequently an Ordinance was promulgated by President on 30 October 1997 making far reaching changes in the Act.

3. International Commission for the Study of Communication Problems (1982)

The Commission, also popularly known as MacBride Commission, was established by UNESCO to study several issues such as increasing importance attached to communication as a social phenomenon and the consequent interest shown in the development of communication media; the growing impact and repercussions of technological progress in the field, etc. The President of the Commission, Sean MacBride, transmitted the Final Report to the Director General, UNESCO in February 1980. The Report is also popularly known as–*Many Voices One World-Communication and Society. Today and Tomorrow.* Some of the important recommendations of the Commission are:

(i) Communication be no longer regarded merely as an incidental service and its development left to chance. Recognition of its potential warrants the formulation by all nations, and particularly developing countries of comprehensive communication policies linked to overall social, cultural, economic and political goals. Such policies should be based on interministerial and inter-disciplinary consultations with broad public participation,

(ii) Within the frame work of national development policies, each country will have to work out its own set of priorities bearing in mind that it will not be possible to move in all directions at the same time. But, as far as resources allow, communication policies should aim at stimulating and encouraging all means of communication,

(iii) The communication component in all development projects should receive adequate financing. So-called development support communications are essential for mobilizing initiatives and providing information required for action in all fields of development–agriculture, health and family planning, education, religion, industry and so on. Essential communication needs to be met include the extension of basic postal services and telecommunication networks through small rural electronic exchanges, and

(iv) Development strategies should incorporate communication policies as an integral part in the diagnosis of needs and in the design and implementation of selected priorities. In this respect communication should be considered a major development resource, a vehicle to ensure real political participation in decision making, a central information base for defining policy options, and an instrument for creating awareness of national priorities.

4. Working Group on Software Planning for Doordarshan (1983)

The Working Group, also popularly known as the Joshi Committee, was established to make recommendations regarding programming in the Indian television. It submitted its report in April 1984, subtitled as *An Indian Personality for Television.* The report was formally laid on the table of Lok Sabha in 1985. Even though the Working Group was formed to make recommendations regarding Doordarshan, its report underlines the overall communication approach and policy concerning other sectors of mass media.

The Working Group Report made several recommendations on creating appropriate environment for the development of creative software for Doordarshan. In the first place, the Working Group was convinced that such an environment did not exist within the present structure of the organization. It was also convinced that whatever the claims of the Government, Doordarshan did not enjoy 'functional freedom' and the lack of such freedom is having a detrimental effect on the planning and quality of its programmes. It recommended the setting up of a National Doordarshan Council which would have three roles to perform : to review and guide the organisations' performance, as a guardian of Doordarshan's functional and professional autonomy, and as a counterpart of the Press Council for examining complaints of inaccuracy or bias in the reporting of news and comments on current affairs. The report emphasized that a developing country like India is exposed to the danger of erosion of its national cultural identity and the need is to resist the cultural invasion from outside through restriction of imported programmes and positive software planning. It discerned a credibility gap created by the growing hiatus between profession and practice, between official policy pronouncements emphasising use of television for development and education, and the increasing drift and departure from them in actual programming. It, therefore, urged the exploitation of the development potential of television of using communication technology for leapfrogging into growth and development. It endorsed the Nehru and Sarabhai approach of tapping the communication revolution in general and television in particular as a major weapon in the fight against poverty and illiteracy.

5. Inter-departmental Committee on Introducing Competition in the Electronic Media (1991)

An Inter-departmental Committee on Introducing Competition in the Electronic Media, under the Chairmanship of K. A. Varadan, Additional Secretary in the Ministry of Information and Broadcasting, was constituted in 1991. Some of the important recommendations of the committee are :

The Committee recognises the importance of introducing competition in the Electronic Media to provide additional outlets to the creative talent, to strengthen the democratic fabric of the nation and help better in meeting the cultural aspirations of different regions. There is need to guard against the dangers of unrestrained commercialism which may follow the establishment of additional channels financed solely through advertising. The dangers of control of these media falling into the hands of unscrupulous or anti-national elements also should not be discounted.

Regional Broadcasts

(i) Subject to viability of operations, preference should be given to setting up new TV/Radio channels in different parts of the country for regional broadcasts. These transmitters can, over a period of time, link up suitably either by forming a consortium or by entering into suitable arrangements amongst themselves. Such a link-up would provide inter-regional network, and

(ii) (a) Simultaneously, the second Doordarshan channel in the four metropolitan cities can be leased out to suitable licencees.

(b) Also simultaneously, additional radio channels including FM stations wherever available may also be similarly leased out.

The option of a licencee leasing a transponder on a satellite for satellite-derived direct broadcast would perhaps be appropriate for developmental or educational telecasts. The Committee notes that as of now there is no such spare space segment available. As and when one becomes available, the Government can take an appropriate decision.

Independent Agency for Granting Licences

The Committee feels that in view of the sensitive nature of the choice, the selection of agencies for grant of licence should be performed by an independent agency in a transparent manner. It feels that the organisations eligible to receive licences should be public limited companies with major interest groups concerned with broadcasting represented in management and with open access to shareholding. Such organisations should be willing to abide by guidelines to be prescribed by the government regarding their functioning. Certain cross media restrictions are recommended in order to avoid monopoly over information sources. Preference may be given to public bodies like universities, cooperative institutions, etc. if any to apply for licences.

Broadcasting Council of India

The Committee has suggested that the Broadcasting Council of India be constituted through a legislation, amending the relevant sections in the Prasar Bharati Act, be entrusted with the functions of licencing, monitoring of programmes and quality rating.

The Committee has recommended that a new legislation be enacted to govern the setting up and operation of additional broadcast channels and has outlined the areas where such legislation will be required.

6. Supreme Court Judgement on Freedom of Airwaves (1995)

The Supreme Court of India, in its judgement, delivered on 9 February 1995 on feedom of airwaves, upheld that airwaves and frequencies were public property. The judgement asked the Central Government to establish an independent, autonomous public authority to regulate their use and lay down important guidelines for Broadcast Law.

While resolving the litigation between the Cricket Association of Bengal (CAB) and Doordarshan over uplinking of self-generated signals to a foreign satellite for viewers abroad, the Court in its trend-setting verdict delivered by a three judge bench headed by Mr. Justice P. B. Sawant said that it was the fundamental right of the telecaster to "educate" the viewers of the game, to watch the telecast was the viewers' sacred right to be entertained. An organiser like the CAB or the BCCI, could not be placed in the same scale as the business organizations wanting to make a large profit by telecasting the series.

Relevant excerpts from the judgement are :

1. The judgement unanimously lays down that, "monopoly over broadcasting whether by government or anybody else, is inconsistent with the free speech, right of the citizens." It is rightly emphasized by Justice Reddy that "state control really means governmental control, which, in turn, means control of the political party or parties in power for the time being. Such control is bound to colour the views, information and opinions conveyed by the media". The unanimous conclusion is that "the broadcasting media should be under the control of the public as distinct from government . . . It should be operated by a public statutory corporation or corporations, as the case may be, whose constitution and composition must be such as to ensure its/their impartiality in political, economic and social matters and on all other public issues. It/they must be required by law to present news, views and opinions in a balanced way ensuring pluralism and diversity of opinions and views. It/they must provide equal access to all the citizens and groups to avail of the medium."

2. Airwaves constitute public property and must be utilised for advancing public good. No individual has a right to utilise them at his choice and pleasure and for purposes of his choice including profit. The right of free speech guaranteed by Article 19 (1) (a)

does not include the right to use airwaves, which are public property. The airwaves can be used by a citizen for the purpose of broadcasting only when allowed to do so by a statute and in accordance with such statute. Airwaves being public property, it is the duty of the state to see that airwaves are so utilised as to advance the free speech right of the citizens which is served by ensuring plurality and diversity of views, opinions and ideas. This is imperative in every democracy where freedom of speech is assured. The free speech right guaranteed to every citizen of this country does not encompass the right to use these airwaves at his choosing. Conceding such a right would be detrimental to the free speech rights of the body of citizens in as much as only the privileged few–powerful economic, commercial and political interests– would come to dominate the media. By manipulating the news, views and information, by indulging in misinformation and disinformation, to suit their commercial or other interests, they would be harming, and not serving, the principle of plurality and diversity of views, news, ideas and opinions. This has been the experience of Italy where a limited right, i.e., at the local level but not at the national level was recognised. It is also not possible to imply or infer a right from the guarantee of free speech which only a few can enjoy.

3. (a) It has been held by this Court–and rightly–that broadcasting media is affected by the free speech right of the citizens guaranteed by Article 19 (1) (a). This is also the view expressed by all the Constitutional Courts whose opinions have been referred to in the body of the judgement. Once this is so, monopoly of this medium (broadcasting media), whether by government or by an individual, body or organisation is unacceptable. Clause (2) of Article 19 does not permit a monopoly in the matter of freedom of speech and expression as is permitted by Clause (6) of Article 19 vis-a-vis the right guaranteed by Article 19 (1) (g).

(b) The right of free speech and expression includes the right to receive and impart information. For ensuring the free speech right of the citizens of this country, it is necessary that the citizens have the benefit of plurality of views and a range of opinions on all public issues. A successful democracy posits an 'aware' citizenry. Diversity of opinions, views, ideas and ideologies is essential to enable the citizens to arrive at informed judgement on all issues touching them. This cannot be provided by a medium controlled by a monopoly whether the monopoly is of the State or any other individual, group or organisation. As a matter of fact, private broadcasting stations may perhaps be more prejudicial to free speech right of the citizens than the government controlled media, as explained in the body of the judgement. The broadcasting media should be under the control of the public as distinct from Government. This is the command implicit in Article 19 (1) (a). It should be operated by a public statutory corporation or corporations, as the case may be, whose constitution and composition must be such as to ensure its/their impartiality in political, economic and social matters and on all other public issues. It/they must be required by law to present news, views and opinions in a balanced way ensuring pluralism and diversity—of opinions and views. It/they must provide equal access to all the citizens and groups to avail of the medium.

4. The Indian Telegraph Act, 1885 is totally inadequate to govern an important medium like the radio and television, i. e., broadcasting media. The Act was intended for an altogether different purpose when it was enacted. This is the result of the law in this country not keeping pace with the technological advances in the field of information and communications. While all the leading democratic countries have enacted laws specifically governing the broadcasting media, the law in this country has stood still, rooted in the Telegraph Act of 1885. Except Section 4 (1) and the definition of telegraph, no other provision of the Act is shown to have any relevance to broadcasting media. It is, therefore, imperative that the Parliament makes a law placing the broadcasting media in the hand of a public/statutory corporate or the corporations as the case may be. This is necessary to safeguard the interests of public and the interests of law as also to avoid uncertainty, confusion and consequent litigation.

7. Working Paper on National Media Policy by the Sub-committee of the Consultative Committee of Information and Broadcasting Ministry (1996)

The Ministry of Information and Broadcasting constituted a Sub-committee of the Consultative Committee of MPs in March 1994 to submit a report on Media Policy. The Sub-committee, headed by Mr. Ram Vilas Paswan submitted its report in March 1996.

The Committee made comprehensive recommendations on all aspects of Media Policy–Print as well as Electronic Media. A summary of important recommendations is as under :

The broadcasting should observe a greater degree of responsibility and sensitivity to Indian culture and ethos and cater to the developmental requirement of the country. It should achieve an Indian personality in telecasting/broadcasting.

There should be a regulatory body to oversee both public and private telecasting/broadcasting. In this connection, the Sub-committee noted that the provisions of the Prasar Bharati Act, which was unanimously passed by Parliament in 1990 should be kept in mind while framing the regulatory mechanism which should be an independent autonomous authority. The Sub-committee is of confirmed opinion that no recourse should be taken by the Government to dilute the provision of the Prasar Bharati Act 1990. This regulatory body may be set up to ensure adherence to the programmes/advertisement codes and other stipulations. This body or similar independent body may also develop an effective mechanism for public grievance redressal as well as implementation of its recommendations in respect of such grievances.

Public Broadcasting

Adequate care should be taken to enable the setting up of non-commercial broadcasting stations to be run by universities, educational institutions, panchayats/local bodies, state governments, etc. The Sub-committee recognises that for public broadcasting, adequate funding in the form of state support is necessary. The content of programmes or software is often dictated by the funding and therefore, the Sub- committee strongly recommends that this aspect should receive the attention of the government and an institutional system be worked out.

The national broadcasters–Akashvani and Doordarshan—should bear the responsibility to offer a high quality public service broadcasting that informs, educates and entertains people and also provides coverage to national events like the Republic Day Parade. Taking into account the reach of the national broadcaster, primary and exclusive responsibility should be provided to them in the matter of coverage of national events like Republic Day Parade, Independence Day, the President's Address to both the Houses of Parliament, etc. The rights/obligations and exclusivety of the national broadcasters should be codified through law.

Whereas the broad policy approach should be the same for radio and television, minor variations/differences can be made in view of the difference in reach and impact of the two media. Programme/advertisement codes should be similar, if not same. The viewer/listener interest should be kept in mind while spacing the advertisements. A new production style, which is people-oriented should be developed. In tune with the policy frame work suggested here, the Indian private sector/state government/NGOs/Local Self-Government should be allowed to enter the field of broadcasting/ telecasting.

Appropriate provisions must be made to ensure that the control of private broadcasting does not fall in the hands of companies having major stake in the print medium or *vice versa.* Cross media ownership restrictions must therefore be considered.

Direct or indirect foreign equity participation in companies entering the field of private broadcasting should not be permitted.

Access on the private channel should be available to the common man through appropriate provisions in the law.

In view of the Supreme Court's Judgement on airwaves, an independent authority must be set up to control and regulate the use of airwaves. The Sub-committee felt that the integrity of the Indian skies is as important as her territorial integrity. This regulatory authority must be set up through a law enacted by the Parliament at the earliest. This body can combine the functions of the regulatory body, or can be a separate body exclusively for controlling the sky waves under the regulatory body.

The apex regulatory body should be an independent autonomous public authority representative of all sections and interests in the society and should control and regulate the use of airwaves in the interests of the public and to prevent invasion of their rights.

The foreign satellite channels must also be brought within the ambit of the said regulatory body in order to make them amenable to the Indian laws. The Sub-committee felt that in the case of skywaves, efforts should be made to develop an international convention to forge multilateral agreements with countries specially in regard to software.

8. Working Group for Reviewing the Provision of Prasar Bharati Act (1996)

As a consequence of advent of satellite channels and their rapid proliferation in the multi-channel scenario, the Ministry of Information and Broadcasting constituted in December 1995 a Working Group under the Chairmanship of Dr. N. K. Sengupta. The Group was entrusted with the task of reviewing the provisions of Prasar Bharati (Broadcasting Corporation of India) Act 1990 and make recommendations regarding its restructuring. The Group, which submitted its report in August 1996, recommended setting up of Radio and Television Authority of India (an independent body which is not a part of Prasar Bharati).

1. Radio and Television Authority of India

Since this constitutes a major recommendation, the relevant excerpts are reproduced as under :

As a sequel to the landmark judgement of the Supreme Court declaring the airwaves as a public property, there is urgent need for establishing a suitable regulatory framework for the multiplicity of channels likely to emerge on the audio-visual landscape in the near future. An authority to grant frequencies and franchises and regulate the licenses has to be created to ensure that there is no unfair or unjust treatment and unwarranted infringement of privacy or violation of accepted standards of public taste and decency. We strongly endorse the recommendation of the Paswan Sub-committee for setting up an independent apex body to regulate and monitor airwaves. We propose setting up a Radio and Television Authority of India, an independent body which is not part of Prasar Bharati. On the contrary, Prasar Bharati being one of the many players in the Indian broadcasting arena, will also be overseen by the Authority as far as the 'complaints function' is concerned. The Authority will perform the following functions :

(i) Licensing of
  - (a) satellite uplinking,
  - (b) terrestrial analog transmission of radio and television,
  - (c) terrestrial digital transmission of radio and television,
  - (d) multiplex providers,

(ii) Prescribing programming obligations and programming standards to ensure quality and diversity,

(iii) Making arrangement for ascertaining and monitoring public opinion about the programmes and services and ascertaining the needs, interests and tastes of members of the public,

(iv) Determine the maximum limit for the quantum of advertising content so that advertisements are not obtrusive or jarring, and

(v) Receive complaints of violation of norms of decency and of the Programming and Advertising Codes and adjudicate on them after hearing the parties concerned.

2. Uplinking Facilities to Domestic and Foreign Channels

We understand that Indian Broadcasters like Jain, Sun TV and Asianet and others like Zee TV, Sony, ATN and BITV have been spending crores of rupees worth of foreign exchange in obtaining uplinking from places like Dubna in Russia, Subic Bay in the Philippines, Singapore and even Sri Lanka. This involved diversion of foreign exchange both by the satellite operators and by Indian companies who have to pay for advertising their products in their channels in foreign exchange. If and when they are allowed to uplink from Indian soil their expense will come down considerably and they will also avoid possible FERA violations. On the other hand, India may gain billions of dollars in foreign exchange through the licence fee and royalty payable by the foreign satellite channels and the advertising revenue that will come from companies outside India. Also this will bring about some real competition for Doordarshan. In the long run this will improve the competitive strength of Doordarshan and will also open up opportunities for news and current affairs programmes on foreign networks. Advertising tariff will substantially escalate as it will be guided by considerations like event marketing and popularity.

Another advantage of extending uplinking facilities to domestic and foreign satellite channels will be that they can be brought under the purview of Indian laws. While granting licences, the Radio and Television Authority of India that has been envisaged, will compel them to adhere to the prevalent Programming and Advertising Codes. The Authority will have the power to prevent telecasting of obscene 'adult' movies and enforce on the satellite channels the same guidelines that DD follows. Similarly, liquor advertisements can also be done away with. Fears have been expressed by some sections about the airing of anti-national programmes or messages on private networks and possible violation of the standards of national security. But we feel such fears are not justified. It is much better to permit and control airwaves than letting them move around as a free floating channel.

3. Prasar Bharati

Prasar Bharati should be a statutory corporation with the preamble :

An Act is to be provided for the establishment of a Broadcasting Corporation of India to be known as Prasar Bharati for providing public service broadcasting and fulfil the basic information, education and entertainment needs of the people and to define its composition, functions and powers and to provide for matters connected therewith or incidental thereto.

4. Public Service Broadcasting

We are convinced that without statutory protection public service broadcasting will be far too vulnerable, especially in a multi-channel scenario. Public service broadcasting needs to be structurally embedded to Prasar Bharati so that it becomes a source of enrichment and empowerment of the masses. The cost of running the public service broadcasting channels, as distinct from purely entertainment-oriented channels, should be met from the Central Government's funding paid to the Corporation by way of grant-in-aid, equity or loan.

5. Local Terrestrial TV and Radio Stations

We recommend that local, terrestrial TV and radio stations should be permitted. These stations will also help municipal bodies and Panchayati Raj Institutions in providing local broadcasting services to the community. For this purpose, the Indian Telegraph Act, 1885 could be suitably amended.

6. Channel for Literacy and Education

Doordarshan and Akashvani should dedicate at least one channel to education, including both literacy promotion and continuing education. Satellite channels including the Open Universities and institutions of higher learning should also be encouraged to do so. The spread of literacy among the people will make them capable of forging their own destinies.

Expansion of Electronic Media

There has been phenomenal expansion of electronic media in the country during the last two decades. Driven by technological developments and demands from an increasingly discriminate and aware segments of the audience, the media scenario in India has undergone a sea change.

Akashvani

Akashvani has three tier system of broadcasting national, regional and local.

The National Channel of All-India Radio started functioning on 18 May 1988. It covers nearly 76 per cent of the population and 64 per cent of the area of the country. It provides a judicious blend of information and entertainment to programmes, including news, high quality Hindustani, Carnatic and Western music, investigative reports and features, magazines, plays, sports, an urdu segment and vividha a causerie of people, men, matters and music.

FM Service as a separate channel is available at All-India Radio Staions in Delhi, Mumbai, Calcutta, Madras and Panaji and time slots are also allotted to private parties for broadcasting programmes.

There are 187 Broadcasting Radio Stations including 72 local Radio Stations and four Relay Centres. The total population covered by medium wave is 97.3 per cent and total area covered is 90 per cent.

Doordarshan

Doordarshan, the National Television Service of India is also one of the largest broadcasting organisations in the world. Ever since it made its debut in 1959 as a modest experimental pilot project, the television world in India today is changing rapidly and fast.

The Doordarshan primary channel operates through a network of 781 terrestrial transmitters, making TV signals available to over 86 per cent of population. Doordarshan programmes are watched by 270 million viewers in their homes. During the years Doordarshan viewership has increased phenomenally and now an estimated 52 million homes have TV sets. India is expected to have nearly 100 million TV homes by the turn of the century.

In 1984, a second channel was introduced in Delhi to provide an alternate viewing option to the metropolitan population. Later this facility was extended to viewers in Bombay, Calcutta and Madras. In 1993 thirty-four terrestrial transmitters were linked through satellite to provide more entertainment for the urban audience. This service–known as Metro Entertainment channel–is now available terrestrially in 43 cities.

Ten Regional Language Satellite Channels were also introduced by Doordarshan in 1995. Several other channels–infotainment channel DD-3, introduced in 1995, has not made much headway. The Movie Club, film channel of Doordarshan is now a part of DD-3.

Cable Television–The New TV Superbazaar

The Special Edition of *India Today,* 1991 caught the spirit of the moment when it heralded the arrival of CNN in January and of STAR TV in April, both in 1991 as "easily the most dramatic revolution took place on the TV screen. 1991 was the year when India ended decades of isolation from the rest of the world." Subsequently in an other cover story, (November 15, 1992), the fortnightly newsmagazine wrote : The mundane television set has suddenly turned into an incredible technicolour dream boat . . . and like a lost tribe discovering an oasis in the desert, lakhs of Indians are hot wired into this superbazaar of soaps, sports, news, music, sex and much more . . Hungry for any kind of news or entertainment, lakhs of people in cities and towns across the nation are now only too familiar with the names of the new electronic media companies, an alphabet soup of initials promising round-the clock nourishment–MTV, CNN, BBC, ATN, Zee TV.

If any thing, the number of satellite channels has increased rapidly. The Sengupta Committee, while referring to Satellite channel universe, lists 52 channels operating in the country.

In order to regulate the operation of the cable television networks in the country, the Cable Television Networks (Regulations) Act, 1995 was passed, but this has not really been given effect to.

Whereas the electronic media have made significant headway in terms of physical expansion, it leaves much to be desired in terms of software planning, objectivity, implementation of the codes and norms for objective news reporting and above all programmes for development in critical areas like agriculture, health, family welfare, etc. That these communication programmes, as far as their actual implementation is concerned, are not commensurate with the policy recommendations and proposals of large number of media committees, is quite evident.

MAJOR POLICY ISSUES

1. Marginalisation of Radio

The Working Group on Autonomy for Akashvani and Doordarshan (1978) had recommended that while television is a powerful and effective medium, there is no doubt that India should give first priority to the development and densification of radio for at least the next 10 to 15 years. Radio is far cheaper and the availability of cheap transistors has made it ubiquitous. The country, however, has a long way to go before it can even begin to think that it has substantially exploited the potential of radio.

Had the above recommendations been implemented, media situation in the country would have been more vibrant, meaningful and relevant to the development needs of the people.

The rapid expansion in the FM radio broadcasting in recent years has brought radio back in the media priorities. Radio is no longer considered as an 'ancient relic' or an attic treasure–relegated to a historical past, preserved for sentimental value rather than for its functional use. The decision of the Government to allot time slots to private producers on FM channels of AIR in the four metropolitan cities of Bombay, Calcutta, Delhi and Madras; putting out live phone-in-programmes; are some of the bold and innovative initiatives which are bound to bring back radio to the centre stage of media world, bustling and booming with excitement and expansion.

Radio for Development

As a tool for development, radio continues to remain unsurpassed because of its easy accessibility, cost effectiveness and localness of the medium. Since television is being driven, more and more by advertising and market forces, the objective of using it as a tool for development has long been relegated to side ways. In fact a stage has reached where even the pretence of the use of television for development is not even being maintained. On television it is soaps, films and film songs, with some programmes on science and environment thrown in. In such a situation it would be once again important to underline the importance of radio for development.

Regarding the role of radio for development, it has been very aptly pointed : Every medium of communication has cultural identity and away from this, it tends to lose its effectiveness. Over the years radio has created for itself an image of being an important source of education and development. This has been possible because of its history of putting out programmes related to social and civic problems, agricultural programmes, health, family planning and nutrition related programmes, and formal and non-formal education broadcasts. The other tele-electronic medium, that is television, on the other hand has evolved for itself an image of being essentially an entertainment medium and even an extended arm of the film industry.

It is, therefore, essential to reiterate that the broadcasting policy be given a critical look keeping in view the primacy that radio should continue to enjoy in terms of accessibility, cost-effectiveness and reach over television for development and educational purposes, whatever might be the entertainment value of the television programmes.

Community Radio

Voices, a Bangalore-based non-profit organisation, an NGO working in the field of media and development, launched a most commendable initiative to promote and strengthen the concept of Community Radio in India. The initiative was the outcome of an International Consultation on Media Policy and Community Radio organised in September 1996 at Bangalore. Over 60 media experts from India and abroad, representing a wide cross section of diverse groups with common concerns in media formulated the Bangalore Declaration on Radio. The Declaration outlines the basic elements of a desirable media policy on radio.

The Consultation highlighted the potential and effectiveness of radio as an essential prerequisite of community-based participatory and interactive broadcasting. Excerpts from the Bangalore Declaration are included as Annexure.

2. Accessibility and Information Gap

Rapid expansion of television during the past decade notwithstanding it still remains inaccessible to a majority of people in rural areas, particularly the poor and the backward, for whom the government wants to disseminate developmental and instructional information. The real potential of the rural audience for Doordarshan programmes remains unexplored. Such a situation is bound to accentuate the knowledge gap between the rural and the urban people with serious socio-political implications.

In a paper : National Cost of Television in India in the Year 2000 : Socio-economic and political implications and issues, Srivastava (1987) has attempted a profile of population having access of TV in the year 2000, which indicates that the knowledge gap between the poor in urban areas and the rich on the one hand and between rural and urban people on the other will be accentuated. It will also have serious socio-economic and political implications.

3. Credibility and Objectivity

Credibility and objectivity of electronic media are the two basic ingredients essential for building up values of trust and reliability between the media and its users. It is more so when the power and reach of the electronic media has increased phenomenally. In order to enforce the values of objectivity and impartiality, various guidelines and codes–The Broadcast Code governing AIR and Guidelines for News Policy for Broadcast Media have been made applicable to the AIR and Doordarshan. However, there has been a visible hiatus between what has been prescribed and what is implemented in practice.

Emphasising the importance of objective and impartial presentation of news, the Verghese Committee observed : The principles of fairness, balance, accuracy and well-rounded objectivity apply not only to news but to all current affairs programmes including commentaries, discussions, news, analyses, documentaries, newsreels and magazine features. Balance, however, does not imply a mechanical balancing of views in each programme, but rather an overall balance achieved over a series of programmes. This is what Akashvani and Doordarshan owe to their audience and this is what the Indian people must always expect of them.

Over the years, due to multiplicity of channels, as also because of opening up of DD II to independent producers–all of them with their news and current affairs programmes, the situation has been mitigated to a large extent. One has only to remember, that under previous dispensation how heads rolled when the name of the daughter-in-law of the Prime Minister got mentioned in the news bulletin in the case of illegal allotment of petrol pump pending before the Delhi High Court. Earlier a full fledged enquiry was ordered as to how a visual of the dozing Prime Minister was included in the news. It appears that Prime Minister and members of his family–their names and visuals–are beyond the pale of normal practice of news production.

4. Development and Electronic Media

The objectives of television broadcasting include : To provide essential information and knowledge in order to stimulate greater agricultural production, to promote and help preserve environmental and ecological balance, to disseminate message of family planning as a measure of population control and family welfare, to promote national integration to act as a catalyst for social change, etc.

Ever since the process of planned development was ushered in the country, with the launching of the First Five Year Plan in 1951, the role of communication in support of development was recognised. The Fourth Plan document stated :

In the spread of information facilities, the imbalance in favour of urban concentrations and prosperous areas continues. There is need for deliberate attempt to inform the people in the rural areas and particularly those in backward regions, about the specific schemes in agriculture, forestry, road construction, marketing, the supply of credit and other inputs so that the benefits of these programmes are more widely spread.

From the Sixth Five Year Plan onwards, funds allocation for electronic media, and particularly for television progressively increased. The Sixth Plan period (1980-85) saw massive expansion of Indian television. The objectives of the Seventh Five Year Plan were :

(a) National and emotional integration of various regions of the country,
(b) Utilization of the medium as a means of development and social uplift, and
(c) Dissemination of information and entertainment.

The Plan envisaged the objective of TV reaching the remotest part of the country through a "three-tier services"–primary service, national service and local service.

Since development, as a subject, has been discussed in greater detail in the chapter on National Development here it would suffice to say; measured against the above stated objectives, Indian television has come no where near achieving these goals. Instead of working as a change agent, TV has ushered in a phase of consumerism that flaunt a highly westernised life style. A sharper thrust, with the greater and direct involvement and participation of the target groups, in the language and milieu familiar to viewers would go a long way in popularising these programmes.

This promising medium has not been able to develop the characteristics of the national broadcasting service, predominently Indian in content and character. Various factors have been responsible for the medium being not closer to the minds and hearts of the people. Its potential to promote the extension of new knowledge and practices, as an aid to transfer of technology, extension education and social change remains unexplored. An autonomous work culture would certainly have encouraged innovation, diversity of opinions and stimulated national consciousness for the disadvantaged sections of the society.

It is obvious that the objectives for which television was introduced in the country, as a medium for social education, weapon against illiteracy and ignorance, for bringing about awareness among the people of sociological problems creating sense of participation in India's efforts to usher in a new social order, for playing a vital role in cultivating civic consciousness; remain, by and large, unfulfilled.

During the past decade the very nature of broadcasting has changed. Broadcasting is no longer a public service, it is becoming more of a commercial venture. The broadcasting industry is becoming increasingly commercial and competitive. Like many other industries, broadcasting is becoming a global business. Increasing number of acquisitions, rise of international networks–MTV and CNN, has made formulation of Electronic Media Policy lot more complicated and complex.

PRINT MEDIA POLICY

The Press in India continues to be in a state of buoyancy in terms of growth in circulation, number of news interest and general newspapers as also the number of languages in which the newspapers are published. Enumerating the salient features of the Press in India, the Registrar of Newspapers for India, in its report for 1995 says :

(i) The total number of news-interest newspapers increased to 35601,
(ii) The total circulation increased to 7,23,02,000 copies in 1994 from 6,36,67,000 copies in 1992, thereby registering an increase of 6.2 per cent,
(iii) Language-wise, the Hindi Press continued to be in the lead both in number as well as in circulation, and
(iv) Uttar Pradesh leads in regard to the number of newspapers as well as in respect of circulation.

During 1994 newspapers were published in as many as 99 languages/dialects. Apart from English and 18 other principal languages enumerated in the Eighth Schedule of the Constitution, newspapers were published in 77 other languages mostly Indian languages or dialects and a few foreign languages. Hindi newspapers numerically constituted the largest group in the country. As in the previous years, highest number of newspapers was published in Hindi (13,650) followed by English, Urdu, Bengali, Marathi, and Tamil language newspapers.

The statistics show that the domination of English language newspapers is receding, more and more regional language papers from non-metropolitan areas are being published. The regional press is increasingly making use of advanced means of telecommunication and printing technology.

Over the past few decades, the Indian Press has witnessed a three tier growth. The main line English language newspapers, published from metropolitan centres, constitute the upper crust of the Indian Press. The English language Press, catering to less than 5 per cent of the readers, has a major share of advertising revenue, because it is seen by advertisers as aimed at the upper class clientele credited with purchasing power.

Leading Hindi and regional language newspapers, which over the past twelve years, have registered a 400 per cent increase in circulation, constitute the second tier of the Indian print media. There has been a spurt of activity in the entire Indian language press resulting not only in vast expansion, but also improvement in quality both editorial and technological. These papers have access to modern printing technology, quite a few of these have multi-city editions published simultaneously from different centres. Several of these bring out colour supplements on week-ends and even on week days.

Small newspapers, coming out from district centres and having grass root contacts, mostly in Hindi and Indian languages, constitute the third tier of the Indian Press. These papers articulate aspirations, demands, difficulties and problems of their readers and have been able to establish strong emotional links with them.

Policy Issues

Some of the major policy issues confronting the Indian Press are : Entry of Foreign Newspapers in India, formulating the code of ethics particularly in the backdrop of the growing corporatisation of media, and reporting communal issues in the language press.

Entry of Foreign Newspapers in India

The entry of foreign newspapers in India is an issue which has polarised the media community, overheated the debate on the subject, with both the protagonists and the opponents speaking on the move with almost religious fervour. Those favouring the move consider the step as the logical outcome of the on-going process of liberalisation, whereas those opposing advise caution against going blind folded into an unknown area and cite the 1955 cabinet decision disallowing such entry as sacrosanct.

In an impassioned article; foreign media inside the Indian fourth estate, V. R. Krishna Iyer, says : To commoditise the media and allow their foreign manufacture within the country is fraught with grave risk to our sovereign democracy and nationalist survival. After all, freedom of the press is but a derivative from the fundamental freedom of expression under Article 19. Freedom of speech, and of occupation or business belong only to the Indian citizens and are subject to reasonable restrictions "in the interests of sovereignty of India. . . friendly relations with foreign states, defamation or incitement to an offence." No 'open seasame' for an outside news heavy weight can be mindlessly granted (*The Hindustan Times,* 11 October 1995).

The Interaction on, Entry of Foreign Newspapers in India, organised by the National Media Centre in September 1994, brought out salient features of the issue in terms of plurality of views, range and presentation of opinion. The former Minister of Information and Broadcasting, P. Upendra suggested that in such a serious matter, which would affect the entire media scene in the country, it should not be left to the government alone to immediately take a decision, a high level committee should be appointed with some government representatives, some media experts, some journalists unions, etc. to go into this problem in-depth and give a report. Ultimately, this must be discussed in Parliament before the government takes a final decision.

Another point of view, articulated forcefully at the Interaction was : Information is on way to becoming the world's largest industry. It will be bigger than the manufacturing of steel or cars and information highways are going to encircle the globe. Technology forces its own changes. Computer technology is converging with telecommunications technology and both are getting together with the information industry. You can soon get video signals on the same optical fibre lines to connect your telephone. You can have data in many forms on the same set of lines. Technology today covers telephones, fax machines, computers, photocopiers and T.V. screens. Encouragement of foreign investment in print media would help India emerge as the hub of information society.

The Roundtable feature in *Media Asia* 22 (4) carried an illuminating discussion on : Foreign Investment in Asian Media– Boon or Bane. In a highly motivated, investment friendly piece, Urban C. Lehner, Editor of *The Asian Wall Street Journal,* argues at length the instance of Singapore making rapid economic headway from 1963-1993, almost a rags-to-rich story, because the key element in its smart economic management was the encouragement of foreign investment. Lehner lists the several advantages of foreign investment in media; infusion of capital and know-how, Asian viewers and readers having more choices at lower prices, foreign investment would lead Asia towards becoming 'information superhighway'. Finally, Lehner condescendingly invokes the Asian governments to recognize that 'foreign investment in Asian media is a boon, not a bane.'

The Chief Editor of *Malayala Manorama* and doyen of the Indian journalists, K. M. Mathew makes out a strong case against foreign investment in print media specially because, whereas laws and conventions in the western democracies place restrictions on foreigners buying stakes in media in their own countries, they would wish India to relax all restrictions and allow anyone and everyone to come to run newspapers in this country.

Those who support entry of foreign investment in media take recourse to the climate of economic liberalisation in the country. If foreign investment is welcome in infrastructure and power industry, why not in print media. Mathew says : We are for liberalisation. I personally and my paper support it. But we want it on our own terms, just as America wants open trade on its own terms. We do not want liberalisation to mean restrictions on our sovereignty and threat to our democratic economy. I am convinced that the entry of all powerful foreign media into India is not in the best interests of Indian media, economy or culture. So, let us be like other democratic countries. Let us not be scapegoats because of our successful democracy. Let us remember that liberalisation in the economic field does not warrant a corresponding action in print media. Liberalisation in newspaper publication is totally different from that of consumer products like edible oil, tooth paste, soft drinks or potato chips. Our media, the sensitive channel of interaction must not be allowed to be swamped by powerful force from outside with no concern for India's interest. Mathew has built up a comprehensive and plausible case against the entry of foreign media in the country.

In a signed article in *Manorma Yearbook* 1998, based on

excerpts from conference papers presented in India and abroad, Mathew says : "Even in several developed countries, that too in the western world, where cultural differences are only marginal, foreigners are not allowed to own newspapers." It is pointed out that even Rupert Murdoch, an Australian, was not allowed to buy a publishing company in the U.S. before he became a U.S. citizen. Even a professional journalist like Harold Evans, ex-Editor of *The Times,* UK could not just walk into a journalistic job in the USA. He had to produce documents before he was admitted. In other words, foreign media barons and even journalists cannot just walk into America and Britain and do what they like, and write what they wish. They have to be "clean" in the eyes of the government; it is good to recall the words of the former National Heritage Secretary of Britain, Mr. David Mellor, reported in the *Guardian Weekly* of September 4, 1994. He said, "One of the great self-inflicted wounds of Britain in the 1980s has been to allow so many national newspapers to fall into the hands of foreign companies who sometimes delight in demonstrating that they have no long-term interest in Britain and its well being." The laws and conventions in those democratic countries put various restrictions on foreign media. . . . Unlike the Indian press, the foreign newspapers are not expected to have any concern for our culture, ethos or tradition. The cultural and moral fallout of the entry of the foreign newspapers would be far more serious than the fallout of the advent of Satellite T.V. in the country where the print media still has greater influence than the electronic media.

Corporatisation of Media

There has been continuing concern over what has been called, "growing corporatisation of the media", in the wake of economic liberalisation. The President, Dr. Shankar Dayal Sharma, while inaugurating the 30th anniversary celebrations of the Press Council of India (PCI) in November 1996 said :

> The recent opening up of the economy and the dramatic expansion of the electronic media have had its impact on the press. Within newspaper networks, there is a trend towards growing corporatisation. How can we ensure that the social objective of informing the public of moulding public opinion and ensuring greater transparency and accountability of those in authority, remains uppermost and that it is not subordinated to commercial consideration.

The Chairman of the PCI, Mr. Justice P. B. Sawant said that unless immediate steps were taken to remove corruption and commercialisation of the profession, journalism would lose its credibility. The PCI has framed guidelines for financial journalists with a view to ensuring a balanced and objective view of financial dealings, status and prospects of a company. The guidelines enumerate, among others that, no newspaper owner, editor or any body connected with a newspaper should use his relations with the newspaper to promote his other business interests.

Several other newspapers, including *Financial Express* and the *Times of India,* have issued code of conduct/ethics for the journalists working in their papers. However, for effective implementation of the Code, it would be necessary to get the cooperation of professional bodies like All-India Working Journalists Federation, Press Guild, as well as business associations like FICCI and CII.

Reporting Communal Issues in the Language Press

Journalism in vernacular languages in India and several other countries has encouraged linguistic, ethnic and sectarian passions and worked to the advantage of fissiparous tendencies. Vernacular newspapers focus on fostering religious and communal divisiveness and emphasise negative aspects of the news.

As a legacy from the prepartition days, when in India more than the English language press, the vernacular press was responsible for aggravating the communal situation, the situation today is no better.

With competing public and political interests, reporting communal riots has always been a complex task—it has become all the more acute in recent times. There have been instances of deliberate and reckless spread of falsehood where media reportage has exacerbated conflict in which interests of the nation as such have suffered. Sadly it is the later which has become more common in significant sections of the press in recent years.

In the related context of reporting Indo-Pakistan affairs also the regional press has been far from moderate and objective. In a seminar on 'Role of Media in a National Crisis' organised by AMIC, Fazal Qureshi Managing Director and the Chief Editor of the Pakistani Press International News Agency said : The national press failed to play the moderating role which it should always play in the larger interests of regional peace and harmony. It is not only the Pakistan press which deserves criticism on this account. The press in India and Pakistan, with notable exception, have always failed to play their role as peacemaker. The events in Kashmir, Punjab and Sind are most unfortunate and the press cannot black out news of events. However, the press should try to act more sensibly, more soberly and more maturely, pleading for sanity on all sides in its news, editorials and commentaries. It should build up public opinion against use of force and put its full weight behind solutions by dialogue.

Regional Press

The doyen of Indian journalists, the late M. Chalapathi Rau wrote : "The future of the India press means essentially the future of the Indian language press, India will have gradually mammoth circulation in all the Indian languages."

There is no doubt that with increase in literacy and in the purchasing power of the people living in slums in the cities, and in the rural areas, there will be significant increase in the circulation of regional language newspapers, not only in India, but in Asia as a whole. In a futuristic study on the development of the press in the year 2000, it has been stated that the present public policy against the growth of a monopoly press and in favour of small and medium sized newspapers is likely to continue. The vast size of the country, multiplicity of languages and cultures and the varying types of regional and local problems also favour the growth of the small and medium sized regional press. With regional development and growth of democratic institutions, a number of newspapers, primarily for the consumption of the literate masses in the rural areas, may also be published.

Changing Values in Journalism

During the past few years, the lowering ethical values in the media have been widely reported. The pernicious practice of accepting envelopes containing gift coupons or hard cash by the press reporters in India at the press conferences where new issues are launched has been widely commented upon.

Writing in *The Tribune* 'Strange Results of Liberalisation', one of the eminent columnists, M. V. Kamath says : What journalists should be concerned about is the corruption of their brethren on a scale that is mind boggling. Giving a graphic account of the process of cash collection is a day in the life of a reporter, Kamath also points to another aspect of the problem–Conformity. "Does that mean there are no honest journalists in the profession. The issue is not honesty but conformity. If the honest journalist refuses to accept the gifts given to him he falls foul of his colleagues. Every one accepts gifts. That is done thing". In this context Nandini Sahai says in *Vidura* (July 1995) : Then there is another class of journalists who are not doing too badly. They call themselves "business journalists", whose primary function is to attend press conferences, specially those aimed at announcing share-issues. They stand to take home quite a neat sum at the end of the month, mostly in the form of gift coupons and gifts. The *Vidura* issue of June 1994, has Dilip Cherian saying, "My own personal experience is that in business journalism the chances of becoming corrupt are much higher". It has also succeeded in creating a breed of journalists who boldly "offer to pay" the newspaper for the privilege of being their "financial correspondent."

Not that the phenomenon of journalists getting rich in the course of news gathering is peculiar to India. In China, a lot of Chinese journalists "write advertisements" in the guise of news after receiving various benefits. In April 1994, the Chinese Communist Party launched a drive to root out the spreading phenomenon of paid news coverage (cheque book journalism) in the country's media organs. A circular put out by the Central Committee's propaganda department calied for, "unremitting efforts by party organisation to educate journalists on press ethics."

Captioned as "China cracks down on cheque book journalism", the above report carried by the *Times of India,* 27 April 1994, referred to the standard practice followed by journalists in covering promotional events which include a small gift, a red paper bag with cash and a banquet after a press conference. The circular further noted : Some correspondents had seriously disrupted the economic order and harmed the interests of the state and people by taking news stories in favour of the sponsoring parties, it said : The practice of demanding and accepting favours for news coverage reached its climax in 1992. A few reporters could not be attracted to a press conference if they were unable to receive benefits.

Similarly, a fiscal ceiling needs to be arrived at regarding the gifts that companies disburse to newsmen. Should they be limited to calendars or diaries? Or should they include products like watches or suitlengths? How far can one go in accepting such freebees—a free trip, a free lunch. Or may be more? This is a grave ethical question again, this needs to be attended to if doubts have to be removed from people's minds regarding the integrity of the media.

It is suggested that a code of ethics needs to be formulated by media men which should be displayed in the reporter's room and on the newsdesk. It should also be published once a year in all newspapers so that readers know the ethical standards to which newspapers subscribe.

Strengthening Press Council of India

It is essential that PCI, in order to be effective should be able to monitor the press throughout the country and not have to wait until an interested party comes up with a complaint. The PCI, it has been suggested quite often, should have monitoring units or branches in different parts of the country to enable it to mobilize pressure of professional organisations and public opinion so essential for countering the insidious aspects of the working of the press in the country.

All these trends are still evolving and we can foresee some of these gathering strength. At the same time it would also be essential to put in place suitable regulatory measures and media monitors, so that growth of pernicious tendencies could be checked.

REFERENCES

Agnihotri, V. K. (ed.) (1995). *Public Policy, Analysis and Design.* Concept Publishing Company, New Delhi.

Akash Bharati : National Broadcast Trust, Vol. I. (1978). Report of the Working Group on Autonomy for Akashvani and Doordarshan. Ministry of Information and Broadcasting, New Delhi.

An Indian Personality for T.V. Vol. I (1985). Report of the Working Group on Software for Doordarshan. Publications Division, New Delhi.

Bhargava, G. S. (ed.). (1991). *Government Media–Autonomy and After.* Concept Publishing Company, New Delhi.

Gupta V. S. (1995). *Third Revolution in Indian Perspective–Contemporary Issues and Themes in Communication.* Concept Publishing Company, New Delhi.

Gupta, V. S. and Vir Bala Aggarwal (1996). *Media Policy and Nation Building–Select Issues and Themes.* Concept Publishing Company, New Delhi.

Lee, John A. R. (1976). Towards Realistic Communication Policies. Recent Trends and Ideas Compiled and Analysed, UNESCO.

Mathew, K. M. (1998). Impact of Foreign Press on Domestic Media in *Manorama Year book* 1998, Kottayam.

Media Asia (1993). Foreign Investmetn in *Asian Media,* Singapore, 22 (4).

Media Asia (1993). Foreign Investmetn in *Asian Media*, Singapore, 22 (4).

Nandini Sahai (1995). Press Institute of India, Journalists for Sale. *Vidura*, 32 (2) : 10-12.

National Media Policy–A Working Paper (March 1996). Submitted by the Sub-committee of the Consultative Committee of the Ministry of Information and Broadcasting, New Delhi.

National Media Policy (1995). A Discussion Paper. Ministry of Information and Broadcasting, New Delhi.

Nihal Singh S. (1992). *Your slip is showing–Indian Press Today*. UBS Publishing, New Delhi.

Press Council of India (1995). *A Guide to Journalistic Ethics*, New Delhi.

Sawant, P. B., Mohan, S. and Reddy, B. P. Jeevan (1995). Supreme Court Judgement in Civil Appeals Nos. 1429-30: Secretary, Ministry of Information and Broadcasting vs Cricket Association of Bengal.

Sawant, P.B. (1997). Press Council–The Conscience Keeper. In : *Mediascape '97*. National Media Centre, New Delhi.

UNESCO (1982). *Many Voices, One World*. The International Commission on the Study of Communication Problems.

5

DEVELOPMENT COMMUNICATION— The Emerging Paradigm

Since the eighties and even earlier, the term development has been defined and redefined several times over, the emphasis shifting from economic growth to distributive justice; and later on development being considered as an inalienable part of the Human Rights' movement. The earlier definition was largely in terms of pursuit of increased production capacity. For a long time development was equated or even held synonymous with economic growth only–devoid of all social implications. It is no longer so. Development has been interpreted in various ways by sociologists, economists and development planners. To some, it means modernisation, to others social change, yet to some others it would mean enhancing quality of life or sustainable development. Development suggests some amount of improvement, self-reliance, uplift and progress; and also self awareness and independence to an extent.

Development has become a catch word, and reference to this term has become the order of the day–for politicians and public men, administrators, planners and economists. It provides an acid test for the success or the failure of the government of the day. But what does it actually mean. Some of the terms used interchangeably either consciously or unconsciously are : Change, growth, progress, modernisation. Likewise, the debate whether development is an objective, process, programme is unending. Economists and social scientists measure development in terms of such abstract indices as gross national product, per capita income and life expectancy, whereas others have stressed the importance of economic impact in accelerating the pace of development, and view development as an all embracing concept, encompassing economic, social, cultural, educational, human face of economic development as well as political aspects of society.

Economists have largely defined development as a steady rise in per capita income. Other indices such as welfare, literacy and life expectancy were presumed to be dependent functions of income. This approach to development is now questioned : greater emphasis is placed on political will to usher in social reforms with emphasis on objectives, such as human aspects of development linked to quality of life, facilitating and increasing participation of people in the decision making process and equitable distribution of national income.

Right to Development

The right to development was included in the Declaration on the Right to Development adopted by the United Nations General Assembly on 4 December 1986. The Declaration states : The right to development is an inalienable human right by virtue of which every human person and all peoples are entitled to participate in, and contribute to, and enjoy economic, social, cultural and political development in which all human rights and fundamental freedoms can be fully realized. The Declaration recognises development to be a comprehensive economic, social, cultural and political process, which aims at constant improvement of the well being of the entire population and of all individuals on the basis of their active, free and meaningful participation in the development process.

Sustainable Development

Likewise in the wake of emerging crisis of environmental deterioration, the concept of Sustainable Development is increasingly becoming the focal issue. Since it is so, let us see what precisely it denotes. The Brundtland Commission defines Sustainable Development as one which meets the needs of the present without compromising the ability of the future generations to meet their own needs. Whereas economic growth remains essential, especially to the hundreds of millions of impoverished people throughout the world, it cannot, however, be at the expense of air we breathe, the water we drink, the land we farm.

The basic new direction, given to the concept of development, initially articulated in the report of the Dag Hammarskjold Foundation Uppsala, *What now? Another Development* is :

Development of every man and woman–of the whole man and woman–and not just the growth of things, which are merely means. Development geared to the satisfaction of needs beginning with the basic needs of the poor who constitute the world's majority; at the same time, development to ensure the humanisation of man by the satisfaction of his needs for expression, creativity, conviviality, and for deciding his own destiny. . . Another development requires transformation of socio-economic and political structures that have long been identified. They include such fundamental steps as agrarian reforms, urban reforms, reforms of the commercial and financial circuits, redistribution of wealth and means of production as well as the redesigning of political institutions through, inter alia, decentralisation with a view to ensuring democratization of the political and economic decision-making power, promoting self-management and curbing the grip of bureaucracies. As history shows, few of these transformation can be achieved without changing the power structure itself.

A widely accepted definition of development was given by Rogers in *Communication and Development* (1976) : Development, according to Rogers, can be described as : A widely participatory process of directed social change in a society, intended to bring about both social and material advancement (including greater equality, freedom and other valued qualities) for the majority of the people through their gaining greater control over their environment.

According to eminent media experts Wilbur Schramm and Daniel Lerner : Development is simply a purposeful change towards a kind of social and economic system that a country decides it wants. The reputed economist, Gunnar Myrdal defines development as 'upward movement of the entire social system.' However, Prof. Cairncross takes a very broad view of development. Development is not just a matter of having plenty of money, nor it is a purely economic phenomenon. It embraces all aspects of social behaviour, the establishment of law and order, the relationships between family, literacy and scrupulousness in business dealings and familiarity with mechanical gadgets.

An other scholar of mass communication has suggested the following definition of development : Development in socio-economic terms means improvement in material welfare especially for persons with the lowest income, the eradication of mass poverty and illiteracy alongwith controlling of disease and early death and change in the composition of inputs and outputs that generally include shifts in the underlying structure of production from agricultural towards industrial activities. In nutshell, it is a process of accelerating the growth with greater social justice aiming at utilizing of resources and opportunities in order to improve the quality of life of the people with special reference to the weaker sections of the society.

Development would mean different things to different people in different countries depending upon the national priorities; one-child family in China, decreased pesticide use in Philippines, creating Computer Culture in Singapore, ORT, clean drinking water and rural roads in India. Development in any particular nation would consist of those projects and programmes which receive top priority in the planning process of that country.

Human development or human rights development has become an important component of the development process. Bella Mody, in her publication–*Designing Messages for Development Communication,* has stressed that development would have to include mobilization and reorganization of the marginalized majority to counteract the continued monopolization of all opportunities for growth by a strong minority.

Mahbubul Haq. in his book, *The Poverty Curtain,* suggested the following salient points in his concept :

- Growth in GNP often does not filter down : what is needed is a direct attack on mass poverty,
- The market mechanism is often distorted by the existing distribution of income and wealth : it is generally an unrealiable guide to setting national objectives,
- Institutional reforms are generally more decisive than appropriate price signals for fashioning relevant development strategies,
- New development strategies must be based on the satisfaction of basic human needs rather than on market demand,
- Development styles should be such as to build development around people rather than people around development,
- Distribution and employment policies must be an integral part of any production plan; it is generally impossible to produce first and distribute later, and
- A drastic restructuring of political and economic power relationships is often required if development is to spread to the vast majority of the population.

The United Nations Development Program's 1996 analysis of the last three decades of development focuses on a comprehensive view of human development. National development must focus on mobilizing human potential through (a) access to material resources like good health, education, housing, and food, and (b) creation of a culture and environment that guarantees freedom, human rights, and personal self-respect. A participatory approach is essential. The United Nations Economic Commission for Africa and a number of citizen groups which met in Arusha in February 1989 explicitly identified lack of popular participation as the primary cause of Africa's crisis. Over centralization of power in the states in Africa had constrained the motivation of its people and underutilized their creativity. The failure of governmental organizations to involve the people and meet their needs has led to the steady growth of non-governmental grass-root organizations.

Development Reconsidered

The *Seminar* (451) in its March 1997 issue on 'Democracy and Development' contains an interesting paper, A New Paradigm, in which environmentalist and writer on economic and political affairs, Sailendra Nath Ghosh, refers to the rethinking and regret on the part of Professor Tinbergen, who propounded the concept of GNP as a measure of progress : The use of GNP is steering the society with a wrong compass. GNP gives, imperfectly and somewhat deceptively, a measure of the aggregate value of national goods and services. It conceals the distribution of wealth and gives a deceptive picture of social health. Since the computation of GNP is not accompanied by a computation of Gross Destruction of Basic Resources (GDBR)–soil degradation, pollution of water and the atmosphere, depletion of mineral resources, the entry of toxic wastes in the food chain, and so on–it is a faulty indicator of economic health. Similarly, balance of payments, in an export-led-growth system gives a highly distorted picture of the internal purchasing power of national currencies.

Ghosh says : Today's tragedy is that mere growth which suits business interests, masquerades as development. In a pursuit of material goods, GNP has become the index of material prosperity and the concept of material goods production, in turn, is now anchored in monocultural practices. An increase in a few select products is being sought at the cost of undermining the production base and even the life support systems. Everything of value has been turned upside down in today's commerce-driven civilisation. In the economic sphere, what was tertiary (trade and services like banking, insurance, market information) has come to claim primary importance, and agriculture, the primary source of creation of real wealth, is now dependent on industry.

During the last 15 years or so, development economics has emphasised that the focus should be on creating conditions that enable human values and social life to flourish. For the social scientists that forms the central theme of development.

The following reference from *Indian Development–Selected Regional Perspective* (1996) adequately sums up the above discussion on development. We must not make the mistake common in some circles–of taking the growth rate of GNP to be the ultimate test of success, and of treating the removal of illiteracy, ill-health, and social deprivation as-at best-possible means to that hallowed end. The first and the most important aspect of Nehru's listing of what we have to do is to make clear that the elimination of illiteracy, ill health and other avoidable deprivations are valuable for their own sake–they are the tasks that we face. The more conventional criteria of economic success (such as a high growth rate, a sound balance of payments, and so forth) are to be valued only as means to deeper ends. It would, therefore, be a mistake to see the development of education, health care, and other basic achievements only or primarily as expansions of human resources– the accumulation of human capital–as if people were just the means of production and not its ultimate end. The bettering of human life does not have to be justified by showing that a person with a better life is also a better producer.

Human Development Report Series–The essays on development– shifting paradigms, different perspectives and naunces of the concept, would remain incomplete without detailed reference to the series, since its inception in 1990. The publication of the series has attempted to put an end to 'the mismeasure of human progress by economic growth alone.' The result of the effort is that more and more political leaders and policy makers in a number of countries have endorsed the idea that the process of development must be people-centred, equally distributed and environmentally and socially sustainable. The central message of Human Development Report 1996 is : There is no automatic link between economic growth and human development, but when these links are forged with policy and determination, they can be mutually reinforcing and economic growth will effectively and rapidly improve human development.

These reports have furnished development planners, academics and journalists with a refreshing approach. The themes of some of these reports have been :

1992–Global Dimensions of Human Development,
1993–Peoples Participation,
1994–New Dimensions of Human Development,
1995–Gender and Human Development,
1996–Economic Growth and Human Development.

The Human Development Report-1996 has drawn attention to the growth that is jobless, ruthless, voiceless and futureless. It considers jobless growth where the overall economy grows but does not expand the opportunities for employment, ruthless growth–where the fruits of economic growth mostly benefit the rich, leaving millions of people struggling in ever-deepening poverty, and voicelss growth–where growth in the economy has not been accompanied by an extension of democracy or empowerment. Voiceless growth can also be growth that gives women only a minor role in an economy's management and direction. Human development, if not engendered is endangered.

The overview to the issues contributing to the Human Development Report–1996, points out some of the most important issues which determine how growth contributes to human development :

- *Equity*–the more equally GNP and economic opportunities are distributed, the more likely that they will be translated into improved human well-being.
- *Job opportunities*–Economic growth is translated into people's lives when they are offered productive and well-paid work. An important way to achieve this is to aim for patterns of growth that are heavily labour intensive.
- *Social spending*–Governments and communities can greatly influence human development by channelling a major part of public revenue into high priority social expenditure–particularly by providing basic social service for all.
- *Gender equality*–Fairer opportunities for women and better access to education, child care, credit and employment contribute to their human development. They also contribute to the human development of other family members and to economic growth. Investing in women's capabilities and empowering them to exercise their choices is the surest way to contribute to economic growth and overall development.

- *Good governance*–when those in power give high priority to the needs of the whole population, and when people participate in decision making at many levels, the links between economic growth and human well-being are likely to be stronger and more durable.
- *An active civil society*–Non-governmental organisations and community groups also play a vital part in enhancing human development. They not only supplement government services, extending them to people and groups who would otherwise remain unserved. They also play a vital advocacy role, mobilising public opinion and community action and helping shape human development priorities.

Development and Public Activism

Public activism is considered a legitimate tool for accelerating social and human development. There have been two refresheringly inspiring books–*India–Economic Development and Social Opportunity*, and *A Field of One's Own, Gender and Land Rights in South Asia,* in which well known economists Jean Dreze and Amartya Sen in the first and Bina Agarwal in the second have pointed out instances, firstly where media have helped draw attention to the social and economic deprivation of the economically poor sections of the society, and secondly the role media can play in influencing public opinion on gender issues. Both these books need to be read carefully, not only by academics and policy makers but also by development planners and social activists.

India's democratic institutions—electoral process, news media, village panchayats provide potential bases for public action in seizing opportunities for social and economic development. While analysing the reasons for the frequent occurrence of famines in China, Sen and Dreze point out that the Chinese government did not wake up to the nature and magnitude of the calamity for quite a long time and the disastrous policies were not substantially revised for three years, while the famine raged on because of the informational failure. The information failure, linked to controlled press, duped not only the public in suppressing information about the famine, but the government was also deluded as well. Whereas in a democratic country, no substantial famine has ever occurred because the government which has to deal with opposition parties, face condemnation from public media, and to go to polls on a regular basis, cannot afford not to take prompt action to avert threatening famine. However, the reach of public criticism can be less effective when the deprivations are less extreme or are of regular occurrence, under nourishment and economic insecurity, lack of school facilities and clean drinking water, may not receive critical attention depending upon the particular nature of politics and journalism in the regions involved.

Likewise Bina Agarwal says : "The leverage that gender progressive groups in bargaining with the state can have, depends on a variety of factors : their ability to sway public opinion or to disrupt administrative functioning (through agitations); the political pressure they can build up in various ways, with implications for voting patterns; the attention they can get for their cause through the media; their ability to deliver programmes which state agencies are unable to deliver by themselves; the existence of pressure for gender sensitivity from international aid agencies, and so on."

Communication

The word communication is derived from the Latin "Communis", which means "to make common, to share, to impart, to transmit." Communication is the basic instinct of man and a social, economic, political and cultural need. Without communication no society can exist, much less develop and survive. For the existence as well as the organisation of every society, communication is a fundamental and vital process. The word communication means not only transmission but also community participation.

Communication maintains and animates life. It is also the motor and expression of social activity and civilization; it leads people and peoples for instinct to inspiration, through variegated processes and system of enquiry, command and control; it creates a common pool of ideas, strengthens the feeling of togetherness through exchange of messages and translates thought into action, reflecting every emotion and need from the humblest tasks of human survival to supreme manifestations of creativity or destruction. Communication integrates knowledge, organization and power and runs as a thread linking the earliest memory of man to his noblest aspirations through constant striving for a better life. *(Many Voices, One World)*

Communication is part of the very fabric of society. It takes place at all levels between peoples and between institutions, from government to people, from people back to government, and through many channels both inter-personal and mediated.

Communication caters to several categories of human and social needs, as indicated in the Report of the Working Group on Autonomy for Akashvani and Doordarshan :

Social : For the articulation of individual and community needs and aspirations; and for education;

Political : In order to create an informed and enlightened public opinion which is the basis of a democratic society; to create awareness within individuals and within society; to mobilise people behind national policies and programmes;

Economic : As a means of extension and technology transfer; as part of the modernization and development process; and for providing market information as well as for marketing and advertising; and

Cultural : As a means of individual and community expression, discovery and enrichment, and recreation.

Like development, communication has also been a subject matter for various definitions and conceptualisations. However, the basic features of communication are :

(a) Communication is a process for transmission of ideas, thoughts and feelings from one person to another;
(b) Communication is persuasive and seeks to obtain desirable response to what is being transmitted; and
(c) Communication is a two-way process, both vertically, and horizontally, in the spirit of 'give and take' or 'send or receive'.

Now naturally, the question : What is relationship of communication with development *per se*?

Development Communication

The concept of Development Communication evolved during the last thirty years, does not elicit a question mark any longer. It has achieved a major breakthrough with substantial amount of thinking going on different aspects–both in theory and in practice. Development communication is primarily purposive communication for planned change. It is intended to promote human development which basically means communication that will help reduce poverty, unemployment and social inequality. It is communication with a social conscience.

According to Nora Quebral : Development Communication is

the art and science of human communication applied to the speedy transformation of a country and the mass of its people from poverty to a dynamic state of economic growth that makes possible greater social equality and the larger fulfilment of the human potential. Nora herself is aware of the possibility of refining this definition even though the basic tenets are unlikely to change (AMIC Study Seminar on Development Communication, 1976).

Development Communication constitutes a fascinating subject for more than one reason. It has always engaged and invited a multi- disciplinary approach because of the vastness of the canvas it covers– Development and Communication—both of which have been subject to changing, varying and different meanings and interpretations. It is because both these terms are intimately related to the human society and its affairs in the broadest sense as such. Just as the modern world is confronted with varying levels, stages and standards of development and is divided into developed and the developing world, with all its conflicts, controversies and struggles; and communicator, media it employs to get across the messages—mass media, inter personal, folk, likewise the concept of Development Communication has also been through various paradigms, models and theories.

Quebral says : The very phrasing of Development Communication indicates its orientation to goals, not to means. Its goals are inherent in the meaning given to development. Some of the more qualitative ones are the enhancement of quality of life, the unfolding of human personality, the humanisation of the individual. The humanistic goals, as well as the multifaceted nature of development (the economic, social, political and cultural aspects) predict a wider scope for the study and practice of Development Communication. It is a broad based concept, to give it a practical shape the primary requirement is communication know-how rather than agricultural know-how. It covers a wide spectrum of the factors involved in development : situation analysis, the involvement of people in the planning and implementation of development activities, motivation, behavioural change, and education and training methods. Communication about improvements in all aspects of rural life, not just agriculture.

Development Communication embraces the whole gamut of communication media : mass media, interpersonal media and extension agents and other communication change agents. Development Communication is not telling people about development programmes, as John Woods, one time Director of Development

Support Communication Services says, which is a part of public relations or public information activity. Just as Development Communication is not public relations job for development, it is not manipulation of mass media either. For mass media tends to be unidirectional, has weak feedback mechanism, is passive and covers an extensive area which percludes effective participation in development process. Since Development Communication aims at adaptation and behavioural changes, it has to be both purposive and persuasive.

The role of mass media in Development Communication will not be those of primary tools but would only be supportive and in conjuction with ground level change agents whose role would be of central importance. The process of Development Communication is more personal and has a strong element of human interaction in it. It aims at motivation and adoption as against merely a conveyor of information which is essentially a function of mass media; and immediate feedback is available through interpersonal communication. The interpersonal communication has the advantages of localness, relevance and proximity about them, as have been laid down by Leslie Sargent.

Researches conducted in many developing countries have shown that the mass media, mainly radio and TV are the best communication sources for creating awareness and interest among the audience regarding a new idea or a message. When it comes to adoption process, interpersonal sources such as extension agents, friends, neighbours, local leaders and others are the most effective communication sources. As all these sources are engaged in accomplishing the goals of development, they fall within the ambit of Development Communication.

Interpersonal Communication

Interpersonal communication is a face-to-face exchange between two or more individuals. The message flow is from one to few individuals, feedback is immediate and plentiful and the messages are relatively high in socio emotional content.

The empirical research has revealed that in a democratic society, the mass media are far less potent than had been thought and, in themselves, are quite unable to affect radical changes of opinions, attitudes and actions on subjects that "matter". At the same time one of the major discoveries of research on mass production is the enduring and powerful influence of interpersonal relations, even in 'mass society'.

Katz points out in his essay, Diffusion-Interpersonal influence, that at first interpersonal influence seemed irrelevant to the students of the media of mass communication. Later on several studies were conducted on the influence of different channels of communication. Taken together, these studies suggested that conversation with other people is generally more influential than the mass media in decisions to accept some new idea or practice; that people who influence others are typically close associates.

Rural sociologists and others have found that the channel that makes for awareness of an innovation is typically a formal one–generally one of the mass media, or an institutionalised source such as a dealer or a salesman, whose job is to spread news of the innovation as quickly as possible to as many people as possible. But extremely few people adopt an innovation on the basis of awareness alone.

Interpersonal influence is a two-way communication, the audience can talk back and convince itself. It is flexible, it is more trust worthy. If a goal of interpersonal communication is to achieve a changed attitude to this end, as a rule, interpersonal communication is more effective than mass communication. Development communication emphasises invariably the results, a response to a message transmitted. It is the quality of response that provides the acid test and constitutes one of the salient features of development communication.

Development Journalism in the Indian Context

Development journalism in India made its debut in 1950s as a result of launching of several development oriented projects through Five Year Plans. There have been several innovative development journalism programmes in the country, the most memorable being the series of articles and features, Our Village Chhatera, in the *Hindustan Times* under the leadership of the doyen of Indian journalists–B. G. Verghese. This series blazed a new trail in the area of development journalism and elicited attention due to the freshness and novelty of approach and the interest it had created in the rural problems of the area.

This one instance is referred to with certain amount of nostalgia, but after that there has not been much concerted effort to place development in proper focus with people's participation as the dominant element (some examples of newspapers taking initiative in promoting farm journalism are given in the Chapter on Rural Development and Media Strategies). Development journalism is often viewed with certain amount of scepticism, for instances it is taken as government propaganda or publicity for the development plans which are heralded with great fan fare. But how often there have been follow up stories on the claims made in the government handouts; how often the statistics are cross checked against the ground situation, how often the benefits actually percolate to the target beneficiaries. Development journalism, more often than not becomes hand out journalism–the claim of the District Magistrate receiving priority over what the target group of beneficiaries actually feel about it.

It is against this type of press coverage of development issues that P. Sainath lodges his protest in his book, *Everybody loves a Good Drought,* which has been on the best sellers list for a long time. Sainath says :

The way the press covers development issues does not help. There have been broadly two tracks in what many call development journalism. The government or official mode and the non-governmental organisation (NGO) or alternative journalism mode. Both have largely failed. The first, already discredited, requires little argument (though there are critical elements in this framework too which are not to be dismissed offhand).

It is about what governments are doing which is important. (Some of the things governments do defy the imagination). For example, there are 2,000 water wells sunk in one drought prone area. The critical element of this mode would be : Were those wells enough? Did they come too late? Should there have been another sort cf well?

But yes, the NGO or alternative mode of development journalism has also failed. Worse, its virtuous halo has deflected critical comment. Somehow, there is often a willing suspension of disbelief when journalists begin to interact with NGOs.

Development theology holds that NGOs stand outside the establishment. They present a credible alternative to it. The majority of NGOs are, alas, deeply integrated with the establishment, with government and with the agenda of their funding bodies.

The current development communication model lays lot more emphasis on participatory approach. Participatory development is, however, more difficult to design and implement than development managed from top by administrators. It requires more social learning, risk taking, coordination skills, organisational leadership, patience and time. Social structural change is interrelated with participatory development and it is evolutionary process and contextual. Development is political and without real political commitment and ideology the participatory model cannot work. Many political leaders and local administrators committed to change have found that in spite of goodwill and efforts, participation is difficult to implement. Not knowing how to manage it, and not anticipating the practical difficulties that follow, their commitment to participation becomes diluted when the inevitable difficulties arise. They did not involve the local people, social and environmental factors in project design and implementation, with the result that the intended results fell far short of expectations.

The general status of Development Communication, either as a subject in the academia or an activity in the government is none too flattering. Addressing a professional body of communication experts in June 1993, the Minister for Information and Broadcasting said that programmes on socio-economic developmental issues produced by All India Radio and Doordarshan had failed to fascinate listeners and viewers. He observed : No matter the amount of money spent or the time allotted to the media some of the programmes are dull and do not have any imagination or concern for the client, the minister said expressing concern at the quality of radio and TV programmes on developmental issues. As the various types of programmes oriented on socio-economic development issues are not imaginative and fascinating, the listener or the viewer simply turns off the knob, as a result the developmental measures do not percolate to the people.

This desperate situation can be remedied only through more training opportunities for media professionals about environment, biodiversity, micro credit system being implemented by organisations like Grammeen Bank, Self Employed Women Association (SEWA) and TAMBULI in the Philippines, which is the call name of a series of community radio stations. Likewise NGO's video strategy, widely involving rural people in Bangladesh, Nepal, Sri Lanka, India has also been effective in creating an environment of participation. Some instances of rural press models promoting improved agricultural practices have been enumerated in an other chapter. However, much more will have to be done to make Development Communication effective and result oriented.

REFERENCES

Agarwal, Bina (1995). *A Field of One's Own—Gender and Land Rights in South Asia.* Cambridge University Press.

Bhuiyan, I. Serajul (1996). Development Communication in the 21st Century. Where has it Been, where is it going? Paper presented at the AMIC Silver Jubilee Seminar, Singapore.

Dreeze, Jean and Amartya Sen (1995). *India : Economic Development and Social Opportunity.* New Delhi, OUP.

Dreeze, Jean and Amartya Sen (1996). *Indian Development-Selected Regional Perspectives.* New Delhi, OUP.

Ghosh Sailendra Nath (1992). A New Paradigm, in *Seminar* (451) New Delhi.

Human Rights (1996). Human Rights Commission, Vienna.

Human Development Report (1996).

Katz, Elihu (1968). Diffusion : Interpersonal influence. *International Encyclopaedia of Social Sciences,* pp. 178-184. The Macmillan Company and The Free Press.

Melkote, Srinivas, R. (1991). *Communication for Development in the Third World—Theory and Practice.* Sage Publications, New Delhi.

Mody, Bella (1991). *Designing Messages for Development Communication.* Sage Publications, New Delhi.

Sainath, P. (1996). *Everybody loves a good drought.*

6

MEDIA STRATEGY FOR RURAL DEVELOPMENT

The World Bank Rural Development Policy Sector Study of 1975 defines rural development as : Rural development is a strategy designed to improve the economic and social life of a specific group of people–the rural poor–and involves extending the benefits of development to the poorest among those who seek a livelihood in rural areas. Rural development, in the broad sense of the term holds the key to all economic progress, industrial or agrarian, urban or rural.

Essentially rural development involves :

1. *Physical development :* communication, transport, sewarage, irrigation, energy, forestry, animal conservation, environment,
2. *Economic development :* agriculture, business, industry,
3. *Social development :* education, culture, sports, political activities, and
4. *Health development :* nutrition, sanitation, community health, prevention of water and air pollution.

Ever since the beginning of the planned economic development, a large number of schemes and programmes were launched primarily to improve the living standards of mass of low income population. However, regarding the effect of rural development programmes, Dutt (1990) has pointed out : Rural development has had the effect of increasing income disparities, even though by a process of trickling down it improves somewhat the standard of living of the poor, has dangerous potentials. The discontent caused thereby and by the lack of social security which often accompanies income disparities, frequently manifests itself in undesirable channels such as religious fundamentalism, regional chauvinism and even secessionism.

Rural Development Planning Approaches

The approach to rural development followed so far has been essentially bureaucratic rather than participatory. Rural development plans have been prepared at the national and state headquarters and passed on to local level offices of the government for execution. Involvement of the local people has been rather limited. Consequently, there has often been a mismatch between what the people wanted and what they got from the Government. Whether it is IRDP, DPAP or any other programme, the dead uniformity with which it has occurred in the name of national planning has proved highly unproductive. As Sardar Tarlok Singh has pointed out, the interest at the higher level has been increasingly to get at quick numerical results which are easy to report, add up and publicise in one form or another with too little attention to the true impact on the conditions of the people.

Decentralised Planning–Urgent Necessity

Participating in the D. T. Lakdawala Memorial Symposium in 1992 on *Decentralised Planning and Panchayati Raj,* the Deputy Chairman of the Planing Commission, Pranab Mukherjee observed : Recently I was going through the district planning documents of a particular district and I found that to implement a developmental outlay of Rs. 8 crores in a year, the actual establishment cost is more than Rs. 16 crores. What type of planning is possible in such a situation? And in certain areas I noticed in that district, at the village level at the lowest level, there are as many as 26 functionaries either fully or partly belonging to both sets of Government, State and Central. And none of them are fully engaged. Now how to coordinate these things and have an integrated approach. And if we had that, perhaps with much less cost we could have achieved much more. So, decentralisation of planning, to my mind, is not merely a cliche or slogan, it is an urgent necessity. From our experiences we have found that we cannot achieve some of our targets in the areas of afforestation, literacy and population control. The involvement of people is the key to the success of these programmes. And surely, the involvement of the people is not possible through the bureaucratic apparatus.

In the United Nations University's Project on Socio-Cultural Development Alternatives in a Changing World (SCA),

Modernization and Development; The Search for Alternative Paradigms, Dube points out : Planning continues to have an elite bias; a small section determines what is good for the society and in the process gives high priority to what is good for itself and its own class interests. Poverty is recognized as the major problem but this recognition has become almost ritualistic. It is the current fashion to speak in the name of the common people; but what they really get are a few crumbs of development. There is a reluctance to carry out meaningful and far reaching structural reforms. Many of the so-called reforms are meaningless and the complexities and loopholes built into them are so many and so diverse that the majority can at best benefit from them only marginally. A substantial part of development activity, despite populist postures, continues to be weighted in favour of the small privileged minority which deftly manipulates the levers of powers.

There is increasing recognition that past development efforts had not fully taken into account social and community needs, cultural values and capabilities as key elements in rural development programmes. However, now there is belated realization that the needs of the people should be at the core of the development process which should be driven by social and environmental concerns rather than economic consideration alone.

Rural development should mean not only an increase in the per capita GNP but greater equality in its composition. This would involve greater emphasis on social services like health, education, family welfare to the poor. Involvement of the local people in schemes of rural development is expected to result in better planning and decision making because the local people have a better awareness of their needs and preferences and fuller information on the conditions and possibilities of their areas. The local communities, with a more intimate knowledge of the inter-dependence of activities at the micro level would be in a better position to develop integrated programmes which would avoid a duplication and produce the maximum impact with minimum cost.

In the D. T. Lakdawala Memorial Symposium, referred to earlier, several speakers pointed out that there has been too much of bureaucratisation in the implementation of development schemes at the district level in the field. The veteran broadcaster Dwipesh Chandra Bhaumik, on the basis of his 30 year experience says : In development stories, our tendency is to take the official version as gospel. What the District Magistrate or the head of the project says has always been more important to us than what the target group beneficiaries feel about it. I am not suggesting that we black out the official version, but if we want our stories to be credible, we cannot afford to ignore the people's point of view . . . Many of us, have been brought up in the mind set of 'two-step' model. We talk down to our listeners instead of talking to them. It is true that radio/ television is essentially a one-way communication. We decide about what to broadcast, when to broadcast without bothering much about the audience reaction. And since the audience has no way of communicating with us, barring the occasional letters, we live in a world of make–believe thinking that whatever we broadcast/telecast is right for them and they will accept it. I think the time has come to change this mind set altogether.

Media Strategy for Rural Development

Media strategy for rural development will have to be essentially geared to the ethos and relevance to the people whom development plans are intended to benefit. The basis of these media strategies should highlight the localness of approach and dissemination of culturally appropriate and user friendly technology. It has been pointed out : There is a general assumption on the part of media professionals that they can disseminate a particular idea without a dialogue with the grass root people because they consider themselves as 'specialists'. A majority of them show little or no respect for the experience and knowledge of the people–the creative potential of the peasants or slum dwellers. They fail to understand that poor people are more suited to depict the issues concerning them and present their own lives. In order to achieve participation, consultation, cooperation and effective mobilization of people, the media system would need to assist in establishing a sense of community and would need a desire to resolve the issues concerning them.

Rural Press

The rural press, as a medium for community communication, with its potential for creating awareness of and articulating people's problems; as a link between the policy makers, development bureaucracy and farmers; as second line of conversion of scientific knowledge into information for utilization in the field situation, imparting adult education and also as a vehicle for reinforcing the literacy habit; is being increasingly recognised as an important and viable means for the promotion of social awareness and national development. Rural press is a promising medium for disseminating information and knowledge about nutrition, health and hygiene, improved farm practices, family planning measures and development plans to those sections of the society where the information is needed most.

The Indian Rural Press Association (IRPA) defines a rural paper as one which devotes at least 40 per cent of its editorial space to agriculture, horticulture, community development, cooperation and Panchayati Raj, etc. A rural paper would essentially be a language paper published from outside the metropolitan and big cities and most probably from district headquarters. It is essentially a small newspaper which, as per the classification of the RNI has a circulation below 15,000 copies. It caters to local issues of immediate importance to the people. According to another definition, a rural newspaper is one that is primarily published from a rural area, devoting at least 50 per cent of its contents to the problems relating to the rural community, in a language akin to the idiom of the rural readers.

Wilbur Schramm in his book, *Mass Media and National Development,* considered rural newspapers as one of the great movers of national development. Such a newpaper gives the neo-literate reason to learn to read better and makes public participation in govermental policies more informed, relevant and meaningful. Apart from broadening the actual educational base of its readers, the rural papers break down the barriers of isolation and provide written material to the villagers who have learnt to read so that they can apply their reading ability to increase their production.

The printed word, particularly in the rural areas, has a fascination of its own. It is a catalyst of change and growth. To improve the recognition, understanding and appreciation of the printed word, specially for those for whom it is not a routine matter, to interact with books and newspapers, is a worthwhile task. Aldous Huxley has said : Every man who knows how to read has the capacity to magnify himself, to multiply the way in which he exists, to make his life full, significant and interesting. The rural newspaper with its simple language and contents closest to the interest of its readers, reinforces this reading habit, day in and day out. Since reading is basic to the understanding of the working of a host of social, economic and political institutions, the rural press by cultivating and reinforcing the reading habit, performs an important role in enhancing social and political awareness.

The International Commission for the Study of Communication

Problems, while summarising its recommendations says : The development of a community press in rural areas and small towns would not only provide print support for economic and social extension activities. This would also facilitate the production of functional literature for neo-literates as well.

Rural press is best suited to act as the second level of conversion of scientific knowledge into information in a usable form for the benefit of the ultimate clientele–the farmers. It is well recognised that by and large, the first conversion of scientific knowledge into information takes place at the level of research institutions. Literature produced at this stage for common consumption can be utilised by the educated and high level functionaries having access to such literature. Such information is further converted, for the second time, in the form of newspapers, pamphlets, newsletters and magazines, and transferred through the government extension agencies to the lower level government functionaries.

Agricultural Journalism Courses

The Research Bulletin No. 35, brought out by the Division of Agricultural Extension, Indian Agricultural Research Institute, points out that readers regarded the articles written by agricultural scientists to be too scientific and technical. Thus, there is a need to have scientists and extension workers as writers who could write without the theoretical bias and have a first hand knowledge of field problems. They will have to acquire techniques of writing technical articles in simple language by using local equivalents for technical terms. To gear the scientists to the task of writing for farmers is a difficult proposition, mainly because of their bias and preference for research paper writing. To improve the situation, it is necessary that scientists be given certain incentives and some short term training in writing for farmers. The short term training courses can be organised by agricultural universities and extension departments. This would help them in writing articles in simple language.

In pursuance of the above recommendations, the author in 1996 conceived and directed a Post-Graduate Diploma Course in Agricultural Journalism at Haryana Agricultural University. The course is essentially a Distance Education Programme for scientists, extension and development workers, working in agricultural universities, research institutes and development departments of the Central and State Governments. The programme has been developed to achieve the following objectives : To acquaint the participants with the role and principles of agricultural journalism and development communication; familiarise them with the process involved in print and electronic media; expose them to practical skills in writing for farmers and farm families, research publications, feature writing and newspapers and preparing radio and television programmes; and to sensitise them to various tasks and skills in public relations and public service campaigns.

Rural Press Development Seminar

In a seminar on Rural Press Development organised by the Asian Mass Communication Research and Information Centre, Singapore in 1986 the participants observed : The need exists for small and rural newspapers, but the rural press must design an appropriate editorial formula that is attractive to the readership in the community. Management must identify the readership and organise itself to reach it. It must also seek the revenue potential and cultivate and exploit the sources fully. Just as readership exists which cannot be served by the urban press, so does advertising potential which cannot be satisfied by the bigger newspapers. The small and rural newspapers must convince themselves that a special clientele exists for them and is waiting to be satisfied. The experience of successful provincial newspapers confirms the view that both readership and advertising potential exist, and this can be exploited to their benefit if newspapers prove they can provide something special and distinct from what the big newspapers are able to.

In China, the *Peasants Daily,* the biggest national agricultural four page newspaper has a circulation of one million. It began publication as a weekly in 1980 with a circulation of 200,000. Each copy costs about 60 paise, a price most rural families can afford. It sells in all China's provinces including Tibet, as well as ten foreign countries. As a rule, every reporter and editor must spend at least two months in the countryside every year during which time they are expected to familiarise themselves with the local peasants, their aspirations and difficulties. The paper engages 300 peasants as regular stringers and about 700 as irregular correspondents. It chooses the most typical letters for publication while referring the rest to relevant organisations. Very often, investigation groups are sent to the countryside if the letters touch upon matters of significance.

It is in recognition of the importance of the rural press that the IPDC has instituted a prize for rural communication with a view to improving communication in all its forms–more particularly local newspapers, radio and television programmes, printing equipment and films–in rural communities and primarily in developing countries.

Reporting Rural Development : Some Constraints

In India reporting rural development has not received the amount of priority and importance in the media that it should have despite a vast majority of the country's population living in rural areas where agriculture is the major income generating source. Reporting development in media is not as pervasive as reporting politics, conflict or even foreign affairs and defence. It is by and large confined to an inside weekly page. And that too by some selected newspapers and magazines. Rural stories rarely get front page treatment or as it is termed the "page one" priority, unless it were a disaster event such as starvation deaths, flood, famine or a policy pronouncement. According to the popular perceptions, rural development reporting is not considered glamorous, or prestigious enough as political reporting. In the reporter's scale of prestige, rural reporting occupies a very low rung. Every young reporter dreams of becoming a political correspondent and moving around in centres of power rather than take to specialization in rural/ agricultural and development reporting. Apart from the elitist bias of the print media and their concentration in urban areas, many journalists are unable to analyse rural issues. The rural sector is so diffused and variegated including agricultural policies and operations, cooperative and development policies, changing village milieu that it cannot be easily packaged into a neat, well defined slot. These and other constraints have relegated rural reporting to a back seat in an otherwise blooming and bustling media scenario in India.

The leading dailies of the Malayalam press–*Malayala Manorama, Mathrubhoomi, Kerala Kaumudi,* etc. have introduced several innovative features for farming community. In an interesting article : Weekly farm feature in the Malayalam press : a unique experiment, in the Media and Rural India, special issue of *Kurukshetra,* R. Hali highlights the yoeman service being rendered by Malayalam press to the farming community.

Malayala Manorama honours the best farmer of Kerala once in two years by awarding the title *Karshaka Sree,* a citation and Rs. 1 lakh. It has won the appreciation of Dr. Norman E. Borlaug, the lone Nobel laureate for his great contribution in eradicating hunger. The selection process undertaken to choose the best farmer also has won all round appreciation and is commended as a model for several award selection processes. The panel of judges for the selection is led by two outstanding farm experts–Dr. M. S. Swaminathan and Dr. V. Kurien, both World Food Prize winners. *Malayala Manorama's* effort to honour the outstanding farmer has motivated the Kerala government and several organisations in the State to recognize the best farmers for their outstanding success.

Radio Broadcasting for Rural Development

Rural broadcasting, and the place it occupies in the canvas of radio programming in India, represents one of the most successful activities of All-India Radio (AIR). The Radio Rural Forum Project, experimented in Poona in 1956, is a significant milestone in the history of rural broadcasting in the country. This experiment was successful in forging a link between three elements involved in the process–the audience, the broadcasting agencies and the specialist agencies.

Another landmark development in the history of rural broadcasting in the country was the setting up of Farm and Home (F&H) units set up at ten AIR stations by the Ministry of Information and Broadcasting, in consultation with the Ministry of Education in 1965. These units arrange broadcast of programmes on agriculture, horticulture, animal husbandry, poultry, fishery, rural entrepreneurship, credit facilities, agricultural marketing and agro-weather conditions. Emphasis is laid on providing, on a continuing basis, hints on modern agricultural practices, information on cultivation of high yielding crops, soil and water management, fertilizer application, plant protection measures, harvesting techniques, storage and marketing of farm produce and farm operation planning according to prevailing and forthcoming weather conditions. The programmes for rural women include information on nutrition, health and hygiene, mother and child care, home management, handicrafts, and on the participation of women in different occupations and economic entrepreneurship.

Encouraged by the response, a decision has been taken to start F&H units in all the existing, as well as new stations. The emphasis in the F&H programmes is on direct presentation, dictated by the need to convey hard core scientific and technical information with quick, timely guidance. According to the 1994-95 Annual Report of the Ministry of Information and Broadcasting, F&H programmes are broadcast daily from all the AIR stations, for a duration of 40-

60 minutes every day.

In the broadcast of rural programmes, AIR can derive greatest satisfaction at having won the confidence of the traditionally conservative farmers and helping them through the use of new and improved techniques to diversify crops, achieve higher production, with obvious economic advantage to themselves and the country (*Indian Broadcasting*, p. 330).

While studying the overall impact of rural broadcasting, the Vidyalankar Committee, appointed in 1963 by the Government of India to study the publicity programmes of the five year plants, observed : 'Wherever radio has reached, its impact has been direct and decisive, radio has undoubtedly been a powerful factor in much of the changes in the political, economic and social outlooks and attitudes, as have come about in rural India in recent years'.

In a study on the Impact of Radio on Farm Women, Rekha Bhagat, a scientist at the IARI, points out that AIR provides a fairly good coverage on diverse topics like child rearing, home management, domestic hygiene, home economy, etc. During the study, farm women were asked their opinion on the impact of radio on the quality of their life. It was seen that education for better living (68.1 per cent) ranked at the top, followed by increase in knowledge (67.5 per cent) and information on home improvement (63.3 per cent). These figures clearly indicate that radio is an important source of education for rural women. It also added to their knowledge and gave happiness as mentioned by 214 (63.3 per cent) respondents. The other important impact of radio was that it gave knowledge on social issues (56.2 per cent), education for health (56.2 per cent) and knowledge on agriculture (53.9 per cent).

Efforts are now being made to use the popularity of radio to start Farm Schools on the AIR programmes for farmers, particularly women. These programmes are aimed to make them aware of community development programmes, help them become better citizens, guide them to develop independent views on important social issues and give information on home science. These programmes are very popular with women; this could be taken advantage of. The popularity of the drama format suggests that the story forms of messages are preferred by farm women, and could, therefore, be used more often. Dramas on family planning, dowry system, early marriage, health and hygiene should be specially written with rural background in mind.

Local Radio Station : A Future Vision

In 1983, All-India Radio broadcast a series of features under the title India–AD 2000. These programmes tried to project a vision of the future, then 17 years away, from the contemporary perspective and experience. The programmes covered different aspects of life and society and one of the features focused on communication. One of the experts interviewed in the programme expressed the definite view that at the end of the 20th century radio would be the dominant medium in our country (Ghosh, 1988).

As a part of the process of bringing media closer to the heads and hearts of the people and to ensure their greater involvement in the media programmes, the concept of Local Radio Stations has proved most effective.

Local radio is comparatively a new concept of broadcasting in India. Each of the stations serving a small area provides utility services and reaches right into the heart of the community, which uses the microphone to reflect and enrich its life. What distinguishes Local Radio from the regional network is its down-to-earth, intimate and uninhibited approach. The programmes of the local radio are area specific. They are flexible and spontaneous enough to enable the station to function as the mouthpiece of the local community. At present there are 187 broadcasting centres including 72 local radio stations.

The Indian experiment in establishing local radio stations began when the AIR ventured into a new phase of broadcasting with the establishment of first local station at Nagercoil in the Kanyakumari District in Tamil Nadu in October 1984.

The comprehensive documentation of the concept, practical utility of local radio for development, expectations and impact of the Local Radio in the *Development Communication Report No. 64,* January 1989, provides valuable insight into the working and future prospects of this unique experiment in the cause of development communication.

The official guidelines on local radio state that since a local station serves a small area, its aim should be "to reach right into the heart of the community of people" it is expected to serve. A local radio station must be flexible and spontaneous to enable itself to function as the mouthpiece of the local community. It should endeavour to use its microphone to reflect and enrich the life of the society. Its programmes should satisfy the local aspirations of the people whom the station serves. In short, local radio should identify itself so completely with the interests of its local population that the heart of the people beats in every pulse of the programmes it broadcasts.

Community Radio–People's Voice

As a tool for development communication, reaching out to the people with relevant messages and usable information, and to encourage people's participation in the 'Open Forum' which it eventually provides, Community Radio is the ideal medium. Whereas in Sri Lanka Mahaweli Community Radio starting in 1981, and in Philippines, Tambuli–a community radio network with 12 stations operating in remote rural areas have been the two successful examples in organising and operating community radio, in India the initiative to conceptualise, and promote community radio was taken by the *Voices* in 1996 (briefly referred to in the Chapter on National Media Policy).

As a communication resource and advocacy group, *Voices*–a Bangalore-based media NGO has been spearheading a movement to examine issues relating to democratisation of media, the right to communicate and access to information and communication technologies particularly by the economically and socially deprived. Its initiative on Community Radio has recently borne fruit in that the Prasar Bharati CEO, S.S. Gill and Board member B.G. Verghese have supported community based programming and the Board has set up a sub-group on community broadcasting. The *Voices* plans to take up a one year pilot project and shall be on air from Chitradurga FM Local Radio Station to gain experience and expertise, before actually setting up a community radio station.

Television and Rural Development Programmes

It was due to the foresight and the pioneering spirit of the visionary space scientist Dr. Vikram Sarabhai and the initiative of Dr. M.S. Swaminathan that an innovative rural television programme, Krishi Darshan, was crystalised in 1967, and has continued since. Krishi Darshan, the oldest, most established and best known TV programme, targeted at rural communities, particularly farmers, has been the subject of several studies and surveys carried out by the IARI and agricultural universities. Started on 26 January 1967, the programme was designed as a pilot rural television project to telecast information for farmers to raise their productivity. Originally telecast for 20 minutes twice a week, it was extended to 30 minutes thrice a week from July 1970. In August 1976, this was increased to four times a week and presently, it is screened for 30 minutes, five nights a week.

In an evaluation to determine the effectiveness of the programme, Krishi Darshan was monitored regularly for one month in November 1986. The programme generally comprised two 8-10 minutes interviews/discussions with farmers, scientists or officials from farm input organisations, a *bhajan* or a group song for 5-6 minutes followed by a short announcement.

In a comprehensive research study entitled, *Comparative Study of Radio and TV Utilisation in Transfer of Farm Technology,* V. P. Chahal identified entertainment to be the main reason for the purchase of radio and TV sets by the majority of farmers. Many were not aware of the name, time and duration of morning and noon slot programmes for farmers on AIR (Delhi and Rohtak) and Krishi Darshan programmes of Doordarshan, Delhi; most of them listened (morning and noon slot programmes) and viewed farm radio and television programmes casually; in almost no cases were notes made or programmes recorded; low listening, passive viewing and feedback behaviour were noted. Normal sources of farm information i.e. extension personnel from grass-roots to district level, farm radio and TV officials or farm scientists were not approached for seeking advice or getting additional information on the recommendations advocated through farm radio and TV programmes. Few made the effort to write to AIR or Doordarshan on farm problems or to participate in their programmes.

Constraints pertaining to the effective utilisation of farm telecast messages from the standpoint of the farmers have been discussed in this study.

According to this research study, the very serious constraints related to effective utilisation of farm radio programmes from the radio owning farmer's view point were :

- Difficulty in making use of theoretical information,
- Lack of field-based programmes,
- Non-availability of extension personnel for consultation on the advice given,
- Lack of education.

In the case of the TV-owning farming community, the problem areas were :

- Lack of field-based programmes,
- Untimely information,
- Inadequate/inappropriate time and frequency of farm telecast (Krishi Darshan),
- Non-availability of extension personnel for consultation.

Another research study on the *Content Analysis of the Krishi Darshan Programme* undertaken by N.N. Joshi of the Haryana Agricultural University suggested certain steps to make the programme more effective. These included :

- The number of talks on animal husbandry and other allied aspects viz., beekeeping, sericulture, mushroom cultivation, marketing, post-harvest technology, need to be increased. This would help in the gradual diversification and commercialisation of agriculture and ultimately in increasing the income of the farmers.
- More emphasis should be placed on group discussions which include farmers, alongwith scientists, and the compere. The presence of farmers would ensure more practical question-answer sessions.
- Only selected visuals of good quality should be included in the programme.
- Training on the preparation and presentation of talks on television should be carried out with special reference to the needs of the farming community.
- Timely telecasts should be ensured particularly for agriculture, where an untimely talk has no practical utility.

The study also recommended that the use of technical and difficult words should be avoided as far as possible; simple and easy-to-understand language would ensure that the content of the telecast is clear, even to the illiterate farmers. Programmes on agricultural themes should preferably be shown in the field situation, which would place the subject in a realistic situation, rather than in the form of a talk in the studio using a specimen or model.

Assessing Rural Television

That the Indian television has become primarily a medium for family entertainment and development programmes do not enjoy a large viewership has also been highlighted by a Doordarshan study. The study conducted by the Audience Research Unit of Doordarshan during January-March 1993, covering 420 villages from 15 states, observed that though Doordarshan entertainment programmes have a large viewership in villages, the development oriented programmes do not enjoy the same popularity. Some of the other findings of the study are :

1. Regional programmes are assumed to serve a very large area. Yet the programmes of Bangalore Kendra may not be relevant to the people in the northern districts of Karnataka. This lack of local specificity and context has a bearing on the viewership. The Doordarshan development programmes have, therefore, not been as effective as they should have been.
2. A total viewership of any programme is dependent not only on the rating, but also on the number of persons viewing from a set. A common observation is that in villages, domestic sets often become community sets, attracting people from other households.

This study also collected information about the average number of persons watching a particular programme from one set–whereas nine viewers watched film-based programmes, only two opted for the programmes on agriculture.

Farming Programmes on Radio and Television–A Comparison

In order to evaluate the effectiveness of the Krishi Darshan programme, a survey in the three villages of Hisar district, the hinterland of Haryana was conducted by two faculty members of the Haryana Agricultural University including the author. The objectives of the study were : to study the TV viewing pattern of rural audience, to ascertain the usefulness of Krishi Darshan programme by the rural community, to identify constraints in viewing the programme, and to undertake content analysis of Krishi Darshan programme.

The results of the study were startling : Krishi Darshan programme, it was surprising to note, even in villages was accorded low priority, and was not watched regularly. The single most important constraint in the way of regular TV viewing was the non-availability of electricity on most days at the time of relay. At times frequent current fluctuation also acted as a deterrent. This factor combined with the fact that the relay time of Krishi Darshan did not coincide with the free leisure time of the farmers hampered regular viewing.

Regarding programme presentation, the respondents mentioned that it was not very appealing as most of the programmes were studio-based monologues with low visual content and appeal. In their view, interviews with farmers in their fields alongwith their opinions and experiences on various practices would be more useful. The respondents expressed the view that the recommendations of specialists on TV were more related to particular areas of specialisation, whereas they were more interested to learn about "packages" which could be adopted. Moreover, they stated that the majority of the recommendations could be adopted by big rather than small farmers.

> As a large majority of the respondents listened to radio programmes related to farming, they felt that they were being apprised about recent practices, more through radio than TV. Radio was noted to be a more credible source of agricultural information even in those households which had a TV set, primarily because the programmes were scheduled at more convenient timings, were regular and most of them were directly concerned with field operations.

The respondents also pointed out that most of the programmes gave the impression that these were mostly studio-based interviews, done with the help of few visuals, mostly charts or written material. Of course, plants, seed samples, seed treatment, method of sowing, etc. were also shown by the scientists who had done their home work. But overall the programmes left a distinct impression that Krishi Darshan was produced, directed and conducted mostly in studios.

Rural reporting on television should be given new dimension and different meaning. The format, presentation and language of Krishi Darshan programme continue to be highly loaded; some times programmes are highly informative, but often charts make the discussion resemble a classroom situation rather than a dialogue. At times, only the written material is presented. The participation of farmers, so essential for any successful communication programme, is minimal. Overall, these programmes leave much to be desired.

The Working Group on Software for Doordarshan observed : Communication can also help to convert the development programmes into a partnership between the experts and the people, each educating and helping the other. The trouble with many development programmes such as Krishi Darshan is that they are produced within the studio, often with urban men in rural garb. It is mostly an urban view of rural problems or a view of problems of urbanised villages. Doordarshan will have to go a long way and play an effective role in presenting rural development programmes.

Television and its Impact on Rural Areas

The study conducted by the Audience Research Unit of Doordarshan, (earlier referred to briefly) states : The television ownership though related to economic prosperity also appears to be dependent on the availability of regional programmes. In Gujarat, which is one of the prosperous states, there are fewer sets perhaps because there were no regional programmes till recently. The Hindi states of Bihar, M. P. and Rajasthan do not have regional service and the number of sets is also very low. However, in Orissa, the regional service has not been able to motivate purchase of TV sets in the rural areas.

There are wide variations in TV ownership patterns : Punjab and Haryana account for 4 per cent rural population but have 13 per cent rural TV sets. Uttar Pradesh, Madhya Pradesh, Bihar and Rajasthan account for 43 per cent population, but only 33 per cent sets ownership. The total viewership of any programme is not only dependent on the rating but also on number of persons viewing from a set. A common observation is that in villages, each domestic set becomes a community set–attracting people from other households. In this study, information was also collected on the average number of persons watching a particular programme from a particular set. The following are the details about the number of persons watching different programmes from a single set :

Feature Film	9	Regional Chitrahaar	10
Chitrahaar	9	Regional serials	6
Serials (Hindi)	4	Regional news	5
News in Hindi	2	Programmes on agriculture	2

Whereas the first survey conducted by Doordarshan in 1993 computed rural TV homes at ten million, the second survey conducted in 1997 showed that the figure has more than doubled to 22 million homes.

Surprisingly and contrary to the general belief, over the years public service messages on TV have started making impact on rural consciousness even in states like Rajasthan. In one of the seminars organised in Jaipur on the impact of cable and satellite television in 1997, it was pointed out that village people were far more aware now about working of the government offices, budget allocation and development works earmarked for their respective areas. They have become conscious about immunisation, marriage age of the girl, family planning, etc.

MEDIA STRATEGY

Demystifying the Media Profession

Knowledge, understanding and expertise in communication practice and profession should not be the exclusive prerogative of the media professionals. This need to be extended through a basic training course to all experts and workers at various levels working in the areas of rural development, health, sanitation and other nation-building activities. A basic appreciation of the elements of communication needs to be imparted through appropriate short term courses. Media specialists, sociologists and development planners should work out inter-disciplinary mechanism for identifying the audience requirement, establishing proper feedback systems and extend guidance in the evaluation of effectiveness of communication programmes.

Training Programmes for Rural Development Functionaries

Training programmes for workers in rural development projects like village level workers in communication methods and those of communication practitioners in rural development would go a long way in creating the much needed mutual appreciation of both the aspects. Since with the expansion of the print media the problems of staff training and recruitment are getting bigger every year, in-house training should be organised by the newspapers, besides formal training in journalism schools. An appreciation course in rural problems, preferably under the aegis of a rural institute or an agricultural university would go a long way in redressing this deficiency. A Consultancy Service for the rural press or for those national newspapers who want to report on rural problems, would fulfil the objective in a large measure. It has also been observed that the personnel working in the rural press lack adequate training required to handle their multi-faceted responsibilities. They lack not only adequate editorial ability, but also the perspective necessary to manage various aspects of a newspaper.

Training would help the rural journalists not only to acquire the perspective of a complete newspaper, but also to evolve an ideal editorial mix. It would be advisable to institute Journalists Fellowship Awards for persons working in the area of rural development reporting and communication. The fellowship would enable the working mediamen to work on a selected project and make an indepth study of the subject. Salary for the fellowship period and travel grant would meet the expenses. Such fellowships had been instituted earlier by the Ministry of Environment, Forests and Wildlife as a part of National Environment Awareness Campaign.

Extension Agency Reinforcement

Radio and TV broadcasting, essentially a part of mass media communication process, share the characteristics and limitations of such a system. In such a system amount of feedback readily available is low, the message flow tends to be one-way and can be used to increase knowledge. Interpersonal medium, on the other hand, tends to be two-way, the amount of feedback readily available is high and is best used in situations calling for attitude formation and change.

Interpersonal medium, which in the context of rural development would include extension workers in the agriculture and community development departments of the state governments, personnel of agricultural agencies like MFALA and agro-industries centres and extension set up of agricultural universities and research institutes, would provide necessary reinforcement and supplement media programmes. Coordination between extension agencies and rural broadcasting will make these programmes more purposive and meaningful. These agencies, can in fact, help disseminate information about media programmes including subjects for discussion through numerous magazines, leaflets, folders they bring out regularly and on special occasions like farmers' fairs and field days, as a part of their extension activities. Programme producers and planners need to be suitably involved in modifying programmes to suit the requirement of the target audience.

Coordination between Government Agencies

There is a greater need for opportunities for interaction between the media men reporting rural development and the policy-makers

(Planning Commission, union ministries and corporate bodies working in the field of Agriculture, Rural Development, National Literacy Mission, Environment, Health and Family Welfare) so that the yawning gap between policy formulation and its implementation could be bridged.

REFERENCES

Anjaneyulu, K. (1989). Scaling Down : Local Radio in India. *Development Communication Report* No. 64, pp. 1-2, Washington, D. C., USAID.

Bhagat, Rekha (1985). Impact of Radio on Farm Women. *Maharashtra Journal of Extn. Education* 4 (1) : 53-58.

Bhaumik, D. C. (1996). Development News in and Electronic Media. *Kurukshetra* XLIV (4-5).

Chahal, V. P. (1992). Comparative Study of Radio and TV Utilization in Transfer of Farm Technology, Haryana Agricultural University, Unpublished.

Doordarshan (1993). Television and its Impact on Rural Areas–A Doordarshan Study. In : (eds.) Dua, M. R. and Gupta, V. S., *Media and Development.* Har-Anand Publications, New Delhi.

Dube, (1992). *Modernization and Development.* United Nations University, Project.

Dutt, R. C. (1990). Rural Development, Some Reflections. *Yojana* 34 (14 & 15), pp. 5-8.

Gupta, V. S. (1987). Rural Press Development in India–Status, factors affecting its growth and future prospects. *Media Asia* 14 (2) : 67-72.

Gupta, V. S. (1989). Case Study of Indian Rural Television. The Krishi Darshan Programme, *Media Asia* 16 (4) : 215-218.

Gupta, V. S. and Rajeshwar Dyal (eds.) (1995). *Rural Press—Problems and Prospects.* Concept Publishing Company, New Delhi.

Gupta, V. S. (1996). Media Strategies for Rural Development. In : *Kurukshetra,* Vol. XLIV, No. 4-5.

Ghosh, Avik (1988). A Future Vision in Continuity and Change—Society, Sesquicentennial Publications, *Times of India.*

Institute of Social Sciences (1994). *Decentralised Planning and Panchayati Raj.* Proceedings of the D. T. Lakdawala Memorial Symposium. Concept Publishing Company, New Delhi.

Joshi, N. N. (1990). Content Analysis of the Krishi Darshan Programme. *Indian Journal of Extn. Education* 16 (1 & 2) : 22-31.

Luthra, H. R. (1986). *Indian Broadcasting.* Publications Division, New Delhi.

Schramm, Wilbur (1964). *Mass Media and National Development.* Stanford C.A. University Press.

UNESCO (1982). *Many Voices, One World.* The International Commission on the Study of Communication Problems.

Voices (1997). Consultation on Media Policy and Community Radio. Workshop Report, Bangalore.

7

MEDIA PRACTICE AND NATIONAL DEVELOPMENT

National Development is a complex and multi-dimensional process–signifying the political will enunciated through properly formulated development policies. Economic and media infrastructures are needed to enhance the capability and provide opportunity to each individual to realize his fullest potential. Human development, with the underlying goal of improved human happiness and quality of life constitutes the core component of Nation Building process.

Just as the UNESCO's constitution's preamble opens with the magnificient words, so often quoted : "Since war begins in the minds of men, it is the minds of men that the defences of peace must be constructed," the process of nation-building must begin from the minds and hearts of the people. This process does not call for a doctrinaire approach–it would centrainly call for an approach based on feelings of goodwill, equity and oneness. Equally important would be to face the challenges of forces countervailing this approach-representing fissiparous tendencies, centrifugal and divisive forces and sectarian outlook. As a mobiliser of public opinion media have a significant role to play in forging national unity and sharpening people's perception about nation and national cohesion.

The Report of the Independent Commission on Population and Quality of Life–*Caring for the Future*–defines some of the factors which make life worth living. The report says : the many factors include : making life more liveable through improved individual and collective health and security; dealing with the scourges of poverty and exclusion; raising the levels of literacy, education and access to needed information; rationalizing production and consumption in terms of what the planet's resources can continue to provide–and bringing fairness and equity to all through better-balanced exploitation and use of these resources. Raising the level of literacy and education has also been one of the major concerns of the research studies and writings of Jean Dreze and Amartya Sen in their two books-*India—Economic Development and Social Opportunity* and *Indian Development–Selected Regional Perspectives.* The two distinguished economists argue : Education is not good only in itself, it is also a necessary condition for economic progress.

> *The elimination of ignorance or illiteracy or remediable poverty of preventable disease and of needless inequalities in opportunities were to be seen as objectives that are valued for their own sake. They expand our freedom to lead the lives we have reason to value, and these elementary capabilities are of importance on their own. These capabilities can and do also contribute much to economic growth and to making the growth process participatory and human capabilities are among the chief means of economic success. We must not make the mistake–common in some circles–of taking the rate of GNP to be the ultimate test of success and of treating the removal of illiteracy, ill health, etc. as at best possible means to that hallowed end . . the elimination of illiteracy, ill health and other avoidable deprivations are valuable for their own sake.*

The media have evinced increasingly greater commitment towards Environment, Sustainable Development and related issues. As environmental problems are becoming serious in many parts of the country, particularly so in urban areas, media are increasingly taking note of urban air quality, loss of biodiversity in flora and fauna, protecting forest cover, etc. Protection of environment is becoming one of the focal issues in the agenda for nation building. Several action groups, grass root movements and NGOs are spear heading campaigns, with constructive assistance from media, to seek policy interventions before big dam construction projects take off the boardrooms. Environmental Reporting, ever since it made its debut in 1972 in the wake of Stockholm Conference has certainly made constructive contribution towards nation building. And finally, the concluding section of the chapter, women empowerment and role of media, Beijing and Beyond, looks at the women issues in media and as to how and whether media have succeeded in the continuing quest of women groups and organisations towards women empowerment.

MEDIA PRACTICE AND NATIONAL UNITY

As the country celebrates 50 years of independence, concepts such as National Unity, National Identity, Sub-nationalism, Regional Nationalism and Regionalism are being debated in the context of India as a nation. There is widespread rethinking, almost an attempt to redefine the concept of National Identity. It is being argued and with greater conviction, in these days of 'Cooperative Federalism', that a country as vast and as pluralistic as India can maintain its unity effectively only through a participant and accommodative mode of politics.

It would, therefore, be appropriate to study the extent to which the traditional definition of Nationalism is still perceived to find acceptance, and whether the extent to which media structures have worked in furtherance of the objectives of forging National Unity, or to the contrary, have worked at cross purposes in times of national crisis.

Nationalism all through the history of mankind, has been one of the most powerful urges that moves a people and it is around this feeling of nationalism that cluster sentiments and traditions and a sense of common living and common purpose. National consciousness is a pre-requisite and a determining factor for national unity. National unity, to a very large extent, would depend upon one's perception and concept of nation.

To the German mind, the word nation conveys fascinating nuances and connotations. It conveys the impression of a great, powerful and highly civilized people with its own state, and the word national is employed only for such lofty concepts as national honour, national unity, the national flag. In Burke's view nation is not merely a multitude of human beings living together, but a historical personality with a specific political and social structure, united by a sort of tacit consent and guided by a small class trained in politics and public service and embodying the national traditions. Frederick Hertz, in his seminal study, *Nationality in History and Politics* (1966), points out to the widespread modern usage of nation which identifies a nation with a people constituted as a state. In his view every state forms a nation and every citizen is a member of the nation. This definition, of course, is an exclusively legal one. Many states were or are composed of different nationalities. The Scots, the Welsh and the Irish regard themselves as nations, though they live in a common state with the English. Therefore, in modern history, the moving force is not the legal concept of nationality but the social and emotional force of national consciousness.

Nationalism, inspite of the recent events in several parts of the world, involving ethnic and ideological conflicts is still universal in its influence and appeal. The nationalist ideal, by and large, in most of the countries is deep and strong. The abiding appeal of nationalism to the spirit of man has to be recognized and respected. The concept of nationalism is not the product of one age, nation, class, party, religion or philosophy, but that each has partly contributed to its growth, partly counteracted it, though in different degrees.

National Cohesiveness

While delineating the goals of national development, Edwin B. Parker and Ali Mohammadi, consider maintaining and promoting national cohesiveness as one of the important objectives of National Development Support Communication and add : One of the prime goals of most developing nations is to create a sense of national identity within a society often composed of disparate ethnic, tribal or linguistic segments. Although militant nationalism leading to hostility between nations can be deplored, few question the value of integrating and interconnecting in cooperative fashion the diverse cultures and interests within a society. Nationalism is preferred to tribalism even by idealists who hope for an end to competitive nationalism in a world in which the sense of family of humankind predominates. *(Communications Policy for National Development,* p. 170.)

In order to achieve national cohesion, diverse elements of a society are interconnected so that they can and do communicate with each other. It has been explained : A communication system linking together the members of a society and permitting continuity through both space and time does not guarantee national cohesiveness. Such a communica-tion system is a necessary, though not sufficient condition. If the messages transmitted over the system lead to mutual adaptation and cooperation among different regions, tribes, ethnic and linguistic groups or classes, then a national identity results from the mutual accommodations. On the other hand, if the messages are part of a process by which one segment of a society is exploiting another, then improved communication may merely expose that pattern of exploitation and hasten a polarization that will destroy hopes for national cohesion.

Attributes of Nationalism

Several tendencies and attitudes have been attributed to nation and nationalism and different elements–cultural, linguistic or even religious– have been emphasised by different parties on different occasions and in different contexts. It is however accepted that, by and large, nationalism represents ardent love of the people, unrelenting fight for its true interests, protection against foreign interference and aggression, safeguarding its traditions and stressing national unity and, solidarity; and of course, subordination of particular interests to those of the nation.

A nation, like the tribal, local and professional groups, is a self-conscious group, but unlike the other groups it comprises millions of people belonging to a variety of sub-cultures speaking different languages and professing different religions, all of whom feel they are one. Nationalism thus involves the will of the people and leads to the formation of a national government. It also implies that interests of the nation are more important than interests of any individual or groups within it.

However, cohesive these groups within a nation may be, these could not be entirely homogeneous. They are bound to be diverse in terms of language, religion, economic and literacy status, etc. Forces of nationalism, in order to succeed the onslaught of divisive forces, have to assert themselves through different media, implementing the national policies and counteracting the forces of disintegration and disunity.

Nationalism or national consciousness is complex and abstract. Nationalism is a set of mind. It is the result of social learning. This feeling of national consciousness could be generated only over a period of time, overcoming a large number of obstacles and barriers, vested interests and countervailing influences.

Having attempted to lay down the broad definition of nationalism, how do we go about forging national unity? Should as important an issue as that of national unity, which encompasses the whole nation, cutting across communities, languages, regions and religions be left to the government policies, or should it be left to the political opinion leaders of the country to espouse the cause of national unity, or would media alone be able to cope with this challenging task which has more to do with the thoughts, aspirations and attitudes of the people than with attainment of merely quantifiable targets designed to fulfil a laid down schedule.

National unity, by definition, is a national objective. It cannot be confined to particular groups or classes of people–whether social, religious, economic, poltical or intellectual. Its objective has necessarily to be a super person or national character which transcends and binds together people of every type, profession, region, creed, caste or race in the country into a single identifiable whole.

In the Indian context there have been occasions, and quite often, when the partisan role of media has resulted in exacerbating the feelings of disharmony, rather than reinforcing good will amongst different sections of society.

Role of Media in Forging National Unity

The Report of the International Commission for the Study of Communication Problems–*Many Voices, One World,* lists two of the main functions of communication as :

> *Socialization :* The provision of a common fund of knowledge which enables people to operate as effective members of the society in which they live and which fosters social cohesion and awareness thereby permitting active involvement in public life.
>
> *Integration :* The provision to all persons, groups and nations of access to the variety of messages which they need in order to know and understand each other and to appreciate other's living conditions, view points and aspirations.

In his book, *Nationalism and Social Communication,* Karl Deutsch points out that presence of ample communication facilities is the single most important feature that makes for the development of nationalism. Deutsch goes to the extent of proposing functional definition of nationality which consists in the ability to communicate more effectively and over a wider range of subjects with members of one large group than with outsiders. Easy accessibility to means of communication facilitates social contacts, travel, knowledge of history and culture. Communications help develop a broad and catholic outlook as opposed to insular and local. Deutsch, calls such people as, *mobilised population of the country,* who are mobilised by intensive communication with others, directly and indirectly.

The frequency and intensity of communication with and within different parts of a country would certainly go a long way in encouraging the forces of national unity at people-to-people level. Likewise communication is seen, as a kind of temperature controlling agent. Through the stimulation of aspirations, "Social temperature" can be raised, and the other way round, by providing explanation, the temperature can be lowered.

In Nigeria, media are viewed as national unifier. In the Nigerian context the mass communicators are expected to : Perform the unique role of aggregating and articulating the interests of the people, giving them leadership, enlightening them and pleading their cause with the government in a plural society and to gather information about the problems, needs and interests of the people, and communicate them to the appropriate decision-making centres. Besides mass media in a pluralistic society should see themselves as the melting pot where various cultures dissolve and merge into one national culture.

In China, mass media alongwith face-to-face communication were effectively used for national integration. The media in China focused popular attention on developing China industrially into a modern state. The media have portrayed regional developments in the context of national progress and unity. Just as television enabled Americans across the country to laugh at the same jokes, the media in China greatly increased the number of terms, news items, stories and personalities that the Chinese people now can share. The end result is a much enhanced national awareness. This enhanced national awareness has mobilized the Chinese people in a way that they are now more ready psychologically to be integrated into the new institutions.

These general conclusions form a part of the Michigan Studies on China, *Communications and National Integration in Communist China.* The study, which is a fascinating account of the use of mass media by the Chinese Communist party for its political and ideological use, points out that the media cannot create national integration by themselves. The study also underlined, the role of radio and films being oral media, as being more adapted to rural audiences, whereas newspapers, journals and books being more suited to urban and educated residents.

Alvin Toffler points out in the *Third Wave* that as second wave societies began to build national economies, a basic shift in public consciousness became evident. The second wave swiftly multiplied the number of people with a stake in the larger world. Bit by bit, psychological horizons expanded. The new mass media increased the amount of information and imagery from far away. Under the impact of these changes localism faded. National consciousness stirred.

Media Practice and National Unity–The Indian Experience

In India the primary objective of the national policy for mass media is to strengthen national unity.

Among the nine objectives laid down for Doordarshan, the first two are to act as catalyst for social change and to promote national integration. Likewise, the principles guiding news policy for broadcast media issued by the Government of India interalia state:

The style and methods of newsreporting should reinforce the fundamental principles on which national policies are based. These fundamental principles include territorial integrity, national integration, secularism, maintenance of public order and upholding the dignity and prestige of Parliament, state legislature and the judiciary.

The statement of objectives, or the charter for the National Broadcast Trust, as envisaged in the Akash Bharati Report, recommended interalia, that the Trust shall :

(i) Uphold the unity of the country and the democratic values enshrined in the constitution,
(ii) Safeguard the citizens right to be informed freely, truthfully and objectively on all matters of public interest, national or international,
(iii) Provide a national broadcasting service, predominantly Indian in content and character,
(iv) Produce and transmit varied programmes designed to awaken, inform, enlighten, educate, entertain and enrich all sections of the people with due regard to the fact that the national broadcast audience consists of a whole series of publics, and
(v) Uphold the secular ideal and promote a spirit of truth and enquiry in the country and among all sections of the people.

The imperative need to propagate a national approach to India's problems, and to develop a national consensus on issues which concern the country as a whole was also underlined.

All these statements of objectives, policies and guidelines, however, do not always translate themselves into ideal programmes and writings furthering the objectives of national unity.

The government control over media has counter-vailing impact on National Unity. The government control and policy framework has an inhibiting and restrictive influence on the quality and nature of programmes put up by the electronic media in the country.

There has been specticism whether the National Programme—broadcast simultaneously from all the TV centres at the Prime Time from 8.30 to 10.30 p.m. claim to attract the national viewership throughout the country. The programme has not been received favourably by several states particularly in the South, some of whom have discontinued the Hindi news; the programme has been generally considered an imposition of Hindi on non-Hindi stations. Since most of the programmes are from Delhi, contribution from other parts of the country are not represented.

The main thrust of Doordarshan is not only the conversion of the South to Hindi and the northern way of life but to obliterate by indirect means the culture of the South. It is a kind of psychological warfare that has been unleashed on the South. All the resources of Government of India are employed by Doordarshan to impose not only Hindi but the superior Northern Culture–(R. Gopale Krishana-*The Illustrated Weekly of India,* December 21-27, 1986).

During the recent years, however, monopoly of Doordarshan has been greatly reduced due to the availability of cable and satellite channels in most parts of the country.

National Integration vs Imperialism

Delhi-Centric View

There have also been murmurs of protest over the Delhi-centric view of India, projecting the problems of Delhi as national problems and the Delhi perspective of issues as national perspective. Even the most insignificant occurrence in Delhi becomes a national event, with extended TV coverage, whereas important events in other parts of the country are either not taken note of, or if noted, recede into a background slot. The result is while Delhi mostly sees itself, the rest of India sees mainly Delhi and occasionally glimpses of India as seen by Delhi. The rich diversity of the cultural and linguistic fabric of the country– Dravidian, Tamil, Bengali, North-Eastern, are represented in the National Programme rarely; more as an exception than as a rule. No wonder TV programmes as telecast through the National Channel have certain artificiality about them.

One of the major grievances against the official media has been projecting the Delhi-centric view of the national issues and lack of appreciation for the diversity and plurality of the cultural heritage of the country as manifested through insufficient coverage given to India south of the Vindhayas.

This writer recently came across the eye opening revelations concerning the problem of national integration in Indonesia in the context of social impact of satellite TV. Indonesia, which had made early investment in domestic satellite system, wanted the system to be used for meeting one of the important policy objectives–National Integration. This was so because problem of regionalism–in a country of 350 ethnic groups, 14 major languages and a large number of separate islands–had to be urgently faced. Interestingly predominance of the Javanese groups, in government and administration further accentuated the pressures of regionalism. Consequently a series of territorial conflicts at the margins of the modern nation, from northern Sumatra to East Timor suggested that boundaries as defined from the centre of government in Java have not always agreed with those seen from the margins. So, what has been viewed as national integration from the centre, has been often seen as imperialism from the periphery (*Media Asia,* 23 (2), p. 102).

That this feeling of alienation in different regions and sections of population in the country is not one time occurrence, but recurring and deep seated was brought out in the feature–Society. The Great Divide– in *Outlook* (27 May 1996). The feature details a number of grievances, whereas some of these seem to be highly exaggerated and coloured like cultural, economic and political colonisation but when a leader like Ramakrishna Hegde with a liberal and pan-Indian outlook, thinks that 'projects which we wanted to start were shifted to a northern state, one is bound to sit up and look for remedial mechanism (political as well as administrative) which should be set up in place to redress these grievances. While ignorance on the part of the northerns about the ethos and cultural traditions of the south may just be due to lack of education and interaction, grievances that national institutions like the National Book Trust and Sahitya Akademi have done little to capture the vibrancy of the literary climate down south must be addressed to. And of course, there is whole lot of validity in the observations that Doordarshan news readers take pains to pronounce names like Francois Mitterand and E. Shevardnedz, they do not put in sufficient efforts to learn the correct pronounciation of South Indian leaders. The only extenuating reason for such sloppiness may be : it is more of a casual attitude of individual newsreader rather than as a result of bias on the part of northerns as such.

National Identity

What should be the national identity of a country as vast as India–with sub-continental dimensions, a population nearing thousand millions ranging from the simple tribals and village rustics (which constitutes the majority) to the neo-rich and urban yuppies; traditional with the bullock cart mind-set and the modern professionals; the best anywhere in the world. In terms of religion, the majority comprising Hindu segment or all inclusive secular representation. One nation or 'innumerable nations'–as Pradip Thomas says in his Keynote Address presented at the Consultation on Communication and Preservation of National Identity in Ho Chi Minh City in October 1996–*AMCB* 27 (2). Some significant observations from Pradip Thomas's thesis on, *Inclusive Vision of National Identity,* a valuable contribution to the subject, are reproduced as under :

An inclusive national identity is built on the fundamental premise that any given nation-state consists of innumerable 'nations', each of whom are heir to continuities in terms of history and culture and to a collective destiny. Apart from recognizable nations, there are also marginalized groups, like grassroots, women and the urban and rural poor whose rights to the nation must also be recognized. While a perfect, wholly inclusive version of national identity might be impossible to formulate, particularly in the context of plural societies, it needs to be reasonably inclusive. It would come as little surprise to know that in a great many countries, national identity is more often than not allied to the group that is dominant–numerically, economically and politically. The right to live by and celebrate national identity is in other words, by no means automatic for all those who reside within the boundaries of a nation-state, tension exists between majorities and minorities, between citizens and non-citizens. This tension is played out in the media and is never really reconciled. In order to circumvent this tension we may need to resurrect and employ a more expansive and substantive understanding of citizenship, meaning the right of every individual living within a nation-state, irrespective of or what he or she is, to have access to its benefits, enjoy its securities and at the same time abide by a culture of rights and responsibilities.

Efforts to forge Indian National Identity along secular lines were earnestly taken up by the Indian National Congress and Jawaharlal Nehru, as a part of the Independence Movement. Nehru maintained again and again that Indian unity could be maintained only on the basis of secularism. Nehru, whose commitment for and conviction in the freedom of press was total, felt deeply concerned that newspapers should spread communalism. He was strongly of the view that newspapers spreading communal hatred should be checked. Whereas he was in favour of freedom of opinion, he did not want newspapers, seeking to increase their circulation to spread communal hatred. He said, "It has become absolutely intolerable that a newspaper should spread utter falsehoods and incite communal passions and make money in the bargain."

It would, however, be appropriate to appraise the role of the press in aggravating communal feelings at a time of national crisis.

Journalism Negating Values of Communal Harmony

During eighties, communalism in India acquired an altogether new dimension. In August 1980, a major riot broke out in Moradabad, resulting in the loss of several hundred lives in the communal riots. Rioting started when the Muslims were offering *Eid* prayers. Most of the people killed in the police firing that occurred, lost their lives while praying. The press tried to create an impression that there was a foreign hand behind the riots. One of the eminent journalists and the editor of the *Times of India,* Giri Lal Jain, went to the extent of alleging in a signed article there was a 'foreign hand' behind the Moradabad riots.

That the electronic media in India, under the control and direction of the government, should increasingly project the image of India as the Hindu India and conceptualize India as a Hindu country has been discussed, in detail, and with conviction by Anand Mitra in his book : *Television and Popular Culture in India–A Study of the Mahabharat.* In the concluding chapter, Mitra states :

> Around these practices of a Hindu-Hindi hegemony, a national identity and a nation is imagined. Doordarshan, as a state controlled medium, is able to represent these practices as the national practices. Moreover, Doordarshan is able to articulate itself with the national practices thus reproducing and circulating a limited set of practices as the preferred practices of the Indian nation. These are the practices that reproduce the image of India as Bharat, where the rule of *dharma,* as embodied in the *Mahabharat* and *Ramayan*, are reproduced as natural, unequivocal and preferred. This is articulated with the emerging Hindu reawakening, where the notion of a non-secular Hindu state is becoming increasingly predominant.

The consequences of projecting India as a Hindu society soon manifested in the turn of events at the end of 1992 in the attack by Karsevaks on the mosque at Ayodhya and subsequent riots in Bombay in March 1993. That the chauvinism which was produced around religious serials like *Mahabharat* and *Ramayan,* and encouraged by the state controlled media should contribute to arousing religious fundamentalism and frenzy in the country, bring us face-to-face with the challenges we are confronted with due to the myopic vision of the political parties and politicians, in quest of vote banks on sectarian and communal basis.

It also needs to be stated that journalism in vernacular languages in several countries has encouraged linguistic, ethnic and sectarian passions and worked to the advantage of fissiparous tendencies. Vernacular newspapers focus on fostering religious and communal divisiveness, emphasize casteism and negative aspects of the news. In India, perhaps, more than the English language press, the vernacular press was responsible for aggravating the communal situation. The Muslim League declaration of August 16, 1946 as the 'Direct Action Day' for pressing the Government for a separate Muslim Pakistan was well publicized by the Urdu Press.

With competing public and political interests, reporting communal riots has always been a complex task–it has become all the more acute in recent times. There have been instances of deliberate and reckless spread of falsehood where media reportage has exacerbated conflict in which interests of the nation as such have suffered.

While censuring four leading Hindi dailies, the Press Council of India in its findings in January 1991 in the Ayodhya case said : Gross irresponsibility and impropriety offending the canons of journalistic ethics to promote mass hysteria on the basis of rumours and speculation. One paper carried the report : 'Muslim doctors of Aligarh Muslim University Hospital are deliberately killing Hindu patients. This report was also brought to the notice of Press Council of India. In this context the widely read columinst, S. Nihal Singh said : This was a national crisis with passions at fever pitch over an issue that had a deep and painful past and a disturbing present in the form of recurring Hindu Muslim riots. How did the press report the crisis? Let us first look at the major newspapers in Hindi and Urdu. They were, as a rule, totally partisan. Much of the reporting on either side was tendentious, some of it highly exaggerated and it seemed that protagonists on each side were only too eager to fan the flames of the communal fire. Some of these erring newspapers later won the rebuke of the Press Council of India. Most of the English newspapers behaved responsibly although in some cases, the stories seemed to be coloured to some extent by their reporters' religion.

While writing on *Media, Ethnicity and National Unity,* in the context of media situation in Sri Lanka, Manik De Silva says : It could be credibly argued that the different approaches of the newspapers depending on the language in which they publish, broadly reflects the thinking and attitudes of the readership they serve. Unfortunately, not enough attention is paid by the managers and owners of the newspaper groups to the need to use their publications as instruments for educating readers on the desirability of ethnic harmony and national unity in a multi-lingual and multi-racial society. The English press has demonstrated a willingness to publish statements and articles advocating ethnic harmony. The Sinhalese and Tamil press, while not remiss in this area, have also tended to carry material that focus on differences, often publishing statements, reports and articles that could rouse communal passions. For example, an English newspaper would be less likely to report that 'Tamil terrorists have massacred Sinhalese villagers', while most Sinhalese papers will have no reservation about presenting it that way. When the 1983 riots erupted, with Tamils in many parts of the country attacked by mobs in what has been generally perceived as a backlash of the separatist war in the northern and eastern provinces, the Tamil Press highlighted the many atrocities that occurred but paid little attention to the many instances when victims of the violence had received protection and hospitality from their Sinhalese neighbours.

In the seminar on, 'Role of Media in a National Crisis' organized by AMIC, Fazal Qureshi, Managing Director and Chief Editor of the Pakistan Press International News Agency said :

> The National Press failed to play the moderating role which it should always play in the larger interests of regional peace and harmony. It is not only the Pakistan press which deserves criticism on this account. The press in India and Pakistan, with notable exceptions, have always failed to play their role as peacemaker. A look at the Indian and Pakistani newspapers will tell a similar story. Pakistan newspapers freely sensationalize the actions of the Indian army in Kashmir and the Punjab. In the Indian newspapers, exaggerated accounts of Pakistani infiltrators and saboteurs helping the secessionists in

Punjab and Kashmir get the same kind of prominent display. In reply Pakistani newspapers blame Indian agents for the law and order situation in Sind. The general reporting trend in both countries is to pick and play up the negative and darker side of the picture. The press in the two countries along with politicians and governments, can be accused of continuing to poison the atmosphere between the two neighbours. It is time we realize the dangers of such action, which is not in the interest of the two countries.

Press Council Guidelines for Reporting Communal Incidents

The Press Council of India, while keeping in view that India is a country with a vastly diverse and differing range of languages, communities, regions and also with varying degree of economic prosperity suggested that the Press, in order to achieve national unity, should emphasize those aspects of the national life which underline the common features of diverse faiths and religions and subordination of sectional interests to national unity and integrity. The Council was, however, emphatic that the writings, which merely set forth or explain the grievances of a community, be it in any field : social, political or economic and are meant to ventilate the grievances of a particular linguistic group of regions should not be considered communal. The Press Council laid down the following guidelines as offending against journalistic propriety and ethics :

1. Distortion or exaggeration of facts or incidents in relation to communal matters or giving currency to unverified rumours, suspicions or inferences as if they were facts and base their comments on them.
2. Employment of intemperate or unrestrained language in the presentation of news or views, even as a piece of literary flourish or for the purpose of rhetoric or emphasis.
3. Encouraging or condoning violence even in the face of provocation as a means of obtaining redress of grievances whether the same be genuine or not.
4. While it is the legitimate function of the press to draw attention to the genuine and legitimate grievances of any community with a view to having the same redressed by all peaceful, legal and legitimate means, it is improper and a breach of journalistic ethics to invent grievances, or to exaggerate real grievances, as these tend to promote communal ill-feeling and accentuate discord.

5. Scurrilous and untrue attacks on communities, or individuals, particularly when this is accompanied by charges, attributed misconduct to them, as due to their or of a particular community or caste.
6. Falsely giving a communal colour to incidents which might occur in which members of different communities happen to be involved.
7. Emphasising matters that are apt to produce communal hatred or ill-will, or fostering feelings of distrust between communities.
8. Publishing alarming news which are in substance untrue or make provocative comments on such news or even otherwise calculated to embitter relations between different communities or regional or linguistic groups.
9. Exaggerating actual happenings to achieve sensationalism and publication of news which adversely affect communal harmony with banner headlines or in distinctive types.
10. Making disrespectful, derogatory or insulting remarks on or reference to the different religions or faiths, or their founders.

Reporting Grievances of the Disadvantaged Sections of Society

The disadvantaged and depressed sections of the society feel that their living conditions, their aspirations and even their success stories do not make news unless something out of the way, violent or even bizarre heppens to them. In the *William Allen White Centennial Seminar* in 1969, Carl Rowen pointed out that the grievances of negroes hardly receive any coverage in the American media unless something volatile happens. He stated : The truth today is that someone who legitimately speaks for thousands of negroes, who articulates their hopes and frustrations, can show up in most American cities and get no better than routine press coverage. But let a negro show up who says : "If you do not do this or that we are going to burn down this dammed town." I guarantee you he will make front page headlines and all the TV shows.

How prophetic his observations are, when viewed in terms of contemporary scenario in India. The media must therefore concern themselves with the grievances, problems and issues impinging upon the depressed and deprived section of the society on continuing basis, and not only when the lid blows off.

The instances of misuse of government controlled electronic media for partisan political purposes have been given at several places in the book. The one instance of such a blatant misuse, with far reaching consequences, stands by itself. In his book–*India : The Siege Within– Challenges to a Nation's Unity,* M. J. Akbar gives graphic details how the official media were pressed into service against Farooq Abdullah in order to justify the conspiracy hatched to remove him from the office of the Chief Minister of Jammu and Kashmir. Akbar writes : Frustrated, the Congress (Indira) now shifted the focus to a familiar charge : that Farooq Abdullah was in league with those determined to destroy India. Sikh extremists were being trained in armed warfare in Kashmir, said Congress leaders (including Mrs. Gandhi and her son). There was never any evidence of this. Even the White Paper which the government published on Punjab in 1984, after the army action, could not prove that Farooq Abdullah had any links with Sikh extremists. But the massive government controlled radio and television spread the propaganda to every corner of the country. An incident on 13 October 1983 during a one-day cricket match between India and the West Indies in Srinagar was blown up into a major scandal. A small group of pro-Pakistan students used the opportunity provided by live television coverage of the match to raise anti-India slogans from a corner of the stands. Dr. Abdullah was livid : Our heads hang in shame over the incident, he told a public meeting at Sopore. But rather than portray him as the man who could challenge and control such forces, the government media made it sound as if the Chief Minister himself was behind the demonstration. The fact that Dr. Abdullah launched an angry tirade against the Jamaat-e-Islami after this was rarely mentioned. Nor could the official media find time to report Farooq Abdullah's repeated commitment to the integrity of India. Farooq Abdullah had to be given a bad name before he could be hanged.

That Farooq Abdullah, with impeccable nationalist credentials and high quality patriotism, who always insisted that he was an Indian first and Indian last should become the victim of highly motivated campaign for ephemeral political gains, was not only a sorry example of misuse of government media but also a serious set back to the cause of national unity.

The voices of dissent, anti establishment and opposition, however, unpalatable these might appear to the powers that be, should be given an equal opportunity to be articulated. An environment of openness, and easy accessibility to the divergent views, a tolerant attitude towards opposing ideologies will help bring down temperatures all around. Cultural alienation of people, linguistic differences or communal conflicts can be a threat to the unity of a country anywhere in the world. At one time Canada's internal crisis over Quebec and heightened bitterness between the French speaking and English speaking Canadians had created the real possibility of national disintegration. The Canadian Prime Minister Pierre Trudeau, fighting to preserve the national unity had warned, "If certain centrifugal tendencies fulfil themselves, we will have permitted this country either to break up or to become so divided that its existence and its ability to act as one nation will have been destroyed."

One can only hope that in the Indian context media do not permit themselves to become a part of centrifugal tendencies be these are linguistic, ethnic or other devisive forces undermining the very basis of National Unity.

LITERACY, EDUCATIONAL TECHNOLOGY AND DISTANCE EDUCATION

Human resources can be enriched by raising and enlarging consciousness and by trained competence and skills through education. Education and media have to be harnessed to bring about changes in attitudes and values–so essential for the task of nation building. Whereas need for expanding available opportunities has been generally stressed differing perceptions about formal versus non-formal education media technology and distance education has enriched the debate about various aspects of education and its role in the society.

Writing in, *Communications Policy for National Development,* Davies in his essay, Multi-Media Education points out : People are the *raison d'etre* of an educational system; indeed, they are the very system itself. Education involves initiating people into worthwhile activities, activities that are thought to be valuable enough to pass on. Media have an essential role in this initiation process, whilst technology (in the dictionary sense of 'knowhow' rather than in the more restricted sense of machines) has an important role in assisting us to identify, design and evaluate worthwhile means and ends.

Davies also cautions against mass communication and more conventional multi-media being perceived as panaceas, universal magical wands that can bring about revolutionary educational change and immediate progress in developing countries. His warning, "If care is not taken, the developing world is likely to be strewn with the sad remains of highly expensive communications equipments, unused and unloved as far as teachers and students are concerned, while education continues in an expanded, but still traditional and perhaps impoverished manner. But it does not have to happen. Success depends upon ensuring that media technology fulfills the needs of learners, teachers, employers and society, rather than dictating what those needs should be," should prove as the appropriate antidote to the current craze of finding technological solution to every problem, particularly so in developing countries like India where elementary education still lacks proper basic schooling facilities in a large part of the rural India–trained teachers, school buildings and even books and blackboards in a number of cases.

The scheme of Operation Blackboard, under the NPE 1986, has been a comprehensive illustration of the deficiencies at the primary school education level. Writing in *Seminar* (436), Caroline Dyer using the example of Gujarat points out : Operation Blackboard arose from the All-India Educational Survey findings which revealed that in 1978, 40 per cent of primary schools had no blackboard; in 1986 almost two-thirds of those schools had 4-5 classes operating simultaneously either in a single room (38%) or in two rooms (32%). Belatedly, it was becoming obvious that school related factors must at least to some extent be responsible for the high drop out rate. In 1986, some 49 per cent of primary school children dropped out before completing the lower primary cycle. Operation Blackboard therefore aimed to set a benchmark norm for lower primary schools : a minimum of two rooms, two teachers and minimum essential teaching learning aids. All existing schools were to be brought up to this level and in future no schools should be sanctioned unless they met the minimum norm requirements.

The importance of literacy as a key factor in capability building, and in enhancing the quality of life of an individual has been long accepted. It would be pertinent to refer to what Aldous Huxley had said in this context : Every man who knows how to read has to magnify himself, to multiply the way in which he exists, to make his life full, significant and interesting. Since literacy is basic to the understanding of the working of a host of social, economic and political institutions, even primary level education performs an important part, particularly in rural area. One cannot enter the new communication and social order if one is not literate. "An illiterate person has hundreds of enemies–epidemics, hunger, disorder and humiliation," said the Soviet Government's statement on elimination of illiteracy in 1920. Illiteracy makes one a victim of false information and false consciousness. Therefore, the key to progress is education which alone can uplift a society.

It will take a long time before literacy and education scenario at the primary school level presents a satisfactory situation, what can still be taken up with certain amount of satisfaction is the belated realization that unless radical changes are initiated in redesigning social policies, including those in basic education and elementary health care, the prospects of raising the general living standards of the population in the country would continue to be none too bright. It is in this context that the efforts of organisations like NOS at school level and IGNOU at higher education level at affording education to the neglected and the disadvantaged sections of the society and their attempts to harness new communication technologies become noteworthy.

The contemporary society is characterised by the swift pace of change which affects our lives in diverse ways : rapid technological development, faster evolution and aggregation of knowledge, rising aspirations of the people towards a more fulfilling quality of life, increasing demand for qualified and trained man power, and exponential growth in the media and communication industry. The harnessing of the potential of communication technologies for educational purposes– not in the restricted sense of structured formal academic system, but by way of opportunities for life long learning available throughout the life span of an individual, has been another notable characteristic of nineties which has brought about a conspicuous shift in the paradigm of education and learning.

Keeping in view the new and larger audiences of learners and participants in the educational process, new content and curriculum changes, education and learning are no longer confined to a standard format of a traditional classroom which offers highly restricted environment. Even during the closing years of twentieth century when expanding educational needs have given rise to different modes of teaching, there are subjects and situations–like music, painting and fine arts–which are still best imparted in the traditional Gurukul and *guru-shishy parampara.* Such traditional modes of learning and teaching have their own advantages–the opportunity of greater and personal interaction between the learner and the teacher, and the greater degree of refinement which these areas possibly demand. These traditional methods of imparting education have been, however, too restricted to cater to the growing demand of industry and professions and number of students who wish to acquire a degree of competence and proficiency in the shortest possible time period in order to meet the requirement of ever competitive scenario.

Catering to Expanding Educational Needs

The educational needs of the population have greatly expanded, both in the developing and developed countries. The educational system should, therefore, be geared to meet the requirement of not only the large numbers of people to be reached, but also in terms of new disciplines, curricula and subjects in tune with demands of industry and economy.

The following observations made in the *Broadcasting for Adult Education* (1972), a UNESCO document, underscores the vast responsibility that rests with the educational planners in formulating workable strategies towards providing educational opportunities for those who missed out in the first instance :

> Many thousands of working young people and adults in many countries are looking for a chance to complete elementary, secondary or higher education–those who were not given the chance to receive formal education during their adolescence because of the shortage of schools; those who have missed their chance, dropping out school or college, because of financial, health, family or other reasons.
>
> Human knowledge is rapidly expanding. In many fields degrees gained years ago are no longer adequate qualifications. The introduction of new technology, new production methods, new techniques advance more rapidly than the setting up of new specializations and new curricula at colleges, universities, and other academic and vocational education establishments. In recent years much attention has been given to the more and more urgent continuing education needs of the graduate force.

Sunset Students–Tryst with Education

Of late several Indian news magazines have brought out news features on 'Sunset students', hair turning grey but with one commitment in life–to pursue higher education. All these students–retired/retiring defence personnel, housewives, in-service staff who wish to attain higher educational qualifications for improving their employment prospects, are pursing higher education through distance education and open learning institutes. For the older generation–bitten by the higher education bug, age is no bar.

The role of the teacher—the key to the success of the educational system–also has to be changed; instead of dispensing information and knowledge he has to guide the students in the acquisition of knowledge from the ever expanding learning resources. All these developments have led to a dramatic expansion in the Open and Distance Education not only in schools, colleges and universities but also in industry and commerce. Besides the United States and Europe, the system of the Distance Education and Open Universities, has made significant progress in the Asian region–more particularly in China, India, Australia, Thailand and Malaysia.

Educational Technology

The Open and Distance Learning System would not have created a viable and lasting niche for itself, within the overall traditional educational system of classroom teaching and university education, but for the extensive use of media and communication technologies.

As technology advances, the ideal of learning, "what you want, when you want, where you want, and at your own pace," becomes possible and life long learning is no longer just a slogan. The report of the International Commission for the Study of Communication Problems–*Many Voices, one World* says : The rapid development of communication in most countries, the expansion of various forms of mass communication and in particular of audiovisual communication, combined with the spread of informatics, open up new horizon and multiply the linkages between education and communication. There is an evident increase of the educational potential of communication. Endowed with a greater educational value, communication generates an "educational environment". While the education system loses the monopoly of education, communication becomes itself a vehicle for and subject for education.

The future of educational media appears extremely promising. In recent years, a series of exciting new technological developments have significantly expanded and enhanced the instructional potential of television. Likewise computer technology has advanced at a remarkable pace, and should play a key role in the future educational expansion. The electronic media offers new dimensions to expand educational opportunities, to improve the quality of instruction, and to increase the cost effectivenss of the present educational system.

Advanced communication technologies such as videotex, teletext, videodisc and computers are being used for education in developed countries. Being expensive and sophisticated, these technological devices are not frequently used in developing countries for educational purpose.

Tele-education, Open and Distance Learning

While considering the application of communication technologies to education, tele-education, by all means, has been considered to be a landmark development. The extensive use of television for educational purposes in Asia in the two most populous countries of the region– China and India, the limitations of being one-way medium notwith-standing, is indicated by two important developments : (1) The largest distance teaching institution of the world, the national multimedia distance learning institution in China, is called the Central Radio and Television University (CRTVU), and (2) One of the largest communications experiments in which television programmes were received by local viewers directly from space– Satellite Instructional Television Experiment (SITE) was conducted in India for a period of one year from 1 August 1975.

The above two instances of application of communication technology, with television playing a pivotal role, are being taken for a detailed discussion, alongwith further developments in tele-education.

Central Radio and Television University (CRTVU)

China is one of the pioneering countries in the world to have exploited television medium for expansion of education. The CRTVU employs broadcast as the main delivery vehicle and supplements it with print, audio-visual media, tutorials, and computerized instruction. Programmes are broadcast directly to learners at home, in the workplace or the study centres located throughout the country. The CRTVU courses follow a multi-media approach. The input of non-print media varies according to the subject area–science and engineering courses have greater proportion of TV programmes, whereas radio is the chief medium for social science course. TV programmes are transmitted by the Central China Television (CCTV) nationally through the microwave network.

The CRTVU–the apex institute, is backed up by the Provincial Television Universities. Academic programmes are offered in a large number of disciplines including engineering, sciences, economics and management : Academic courses are designed on a nation wide basis, but provincial universities implement the guidelines provided by the CRTVU. Enrolment and examinations are controlled centrally and most eminent writers and presenters are invited to produce TV programmes.

The remarkable aspect of the working of CRTVU is the expansion of its operations in rural areas in order to meet the shortages of technically trained manpower. In his paper–China's Radio and Television University–World's Largest Network of Open Universities, G. Ram Reddy gives his assessment about CRTVU's contribution to education in China: A vast network has been created to meet the shortage of educational persons in various fields and also to meet the demands of fresh graduates from high schools. Since it is not possible for the provincial universities to meet these demands, the society is looking to the Radio and Television Universities to cater to these demands. During the last eight years these universities have made a tremendous impact on the educational system in the country.

Satellite transmission is the kernel of China's distance university system, linking the Central Chinese Radio and Television University in Beijing to the network of 43 other open universities throughout the country (Wei and Tong, 1994, quoted in the paper by Desmond Keegan).

Satellite Instructional Television Experiment (SITE) 1975—The Indian Experience

The SITE was a one-year pilot project, launched during 1975-76 to assess the feasibility for centrally conceived and produced messages to be distributed simultaneously throughout large geographical areas without relying on extensive relay or retransmission facilities. Such a capability vastly increases the potential audience for development broadcasts of all kinds. Conceivably satellite may soon permit nations to circumvent the development of extensive terrestrial communication systems.

Primarily undertaken to experiment with television through satellite communication to broadcast special programmes in six states which included 2400 villages, the experiment aroused wide international interest. India was the first developing country in the region to undertake an experiment using space age technology to bring its millions into the mainstream of national life, make them aware of the fast changing pattern of present day urban life and help them to harness their resources for optimum benefits. Explaining the importance of SITE, Narayana Menon (1976), in his book, *The Communication Revolution* wrote :

SITE means a lot of things apart from being an experiment.

For one thing, it made India TV-conscious and of an advanced style of TV technology. TV is no more confined to the main metropolitan centres. It has suddenly made its presence felt in the villages of Karnataka, Andhra, Orissa, Madhya Pradesh, Bihar and Rajasthan. Hundreds of engineers, technicians and mechanics are acqring new skills and vast experience. Programmers, broadcasters and telecasters are getting new insights into the medium. In spite of all the doubts and hesitation, bureaucratic stumbling blocks, geneteel in-fighting, shortage of equipment, facilities, money, we are in the middle of a vast learning process, the culmination of the tentative approaches of the tele-club days of 1959. All these things add up to substantial investments which should, in the long run, bring rich dividends. As Vikram Sarabhai said, television in an investment rather than an overhead.

Another observation on SITE is equally noteworthy : SITE experience does provide a vision of what rural communications might become and a much clearer indication than has heretofore been available of the range of administrative and programming that are embodied in so powerful a technology.

Countrywide Classroom (CWCR)

The Countrywide Classroom (CWCR) project of the University Grants Commission (UGC) has been the logical outcome of the successful SITE experience. The rapid expansion of television in India during 1984–when one TV transmitter was being installed every day–facilitated the launching of the above programme, aimed primarily for the undergraduate students of colleges located at different places. The contents of CWCR programmes are not linked to any examination, these are neither based on a particular syllabus. Instead these telecasts seek to provide new insights, bring new findings and take the students to various places and laboratories they would rarely see. The project aims at the widest possible utilisation of television potential to improve the quality of university level education. These educational video programmes are telecast twice a day over the Doordarshan network using INSAT Satellite system. The UGC has established a number of media production centres all over the country for producing these programmes. India, in fact, has one of the largest networks for producing quality education software which include Audio-Visual Research Centres

(AVRCs) and Educational Media Research Centres (EMRCs)–under the aegis of the UGC. The establishment of the Indira Gandhi National Open University represents the government's attempt to provide higher education through the use of multimedia distance teaching methods. The establishment of the Central Institute of Educational Technology (CIET) in 1984 and also the State Institutes of Education Technology (SIETs), with the objective of utilising educational technology for the improvement and spread of education in the country at the school level, is one major step towards utilising the media for elementary education. Some of the major concerns at present–qualitative improve-ment and regular availability of programmes, wider coverage to the users through telecast–have been articulated on different occasions.

Tele-education Programmes of IGNOU

Since 20 May 1991, Doordarshan has been telecasting curriculum based video programmes produced by IGNOU, an important component of its educational methodology which aims at making the teaching-learning process interesting both for the student and the teacher. The target audience for these telecasts are primarily IGNOU students spread over throughout the country.

In order to assess the utilisation of telecasts by the students, as also to get insight into the quality of these programmes, as assessed by the students, a study was conducted by the Communication Division of the University. The study, essentially a one-shot survey research project, was conducted at the 33 study centres located in 16 Regional Centres throughout the country with 431 out of 1498 respondents responding to the questionnaire. The research study was, therefore, at best, sketchy in terms of the total number of courses and the percentage of students who were covered in this survey. The questionnaire was bilingual–in English and Hindi and the participants were drawn from two programmes–Management Programmes and the Bachelor's Degree Programme. The survey revealed that single most important reason for not watching the telecast was unsuitability of the telecast time, the other being non-possession of TV sets. A majority of non-watching respondents indicated that a repeat telecast in the evening would help a large number of students to watch the telecasts. If it was not possible to shift the telecast time or repeat telecast is not possible, some alternative channel of reaching the students would have to be explored.

The study made two important suggestions for strengthening the tele-education programmes of IGNOU :

1. The possibility of telecasting from various regional transmitters with different schedules should be considered in addition to the present nation-wide telecast. Since the mandate of the university is to bring higher education to the large number of citizens, an effort is urgently required to produce programmes in the language the students are adequately conversant with.
2. Close interaction and cooperation among the producers, script writers, faculty members and the production team is required for preparing an effective educational programme.

Educational Teleconferencing

Limitations inherent in tele-education have been pointed out in several discussions on the use of television for teaching and distance education purposes. The most serious limitation, of course, is that television is strictly a one-way communication. The teacher-taught contact, considered all important for any successful educational endeavour is missing in the case of television. Gopal Saksena elaborates this rather graphically : No personal contact, no physical identification, no scope for repetition, clarification or elaboration. The gap between the 'sender' and the 'receiver' is too wide to be easily bridged. A teacher's personal touch and care are not a diminishing commodity subject to the vagaries or showmanship of the medium of television. Hence, TV and teacher should prove to be more of companions than competitors. Reddy points out. The problem with television programmes is that these are ephemeral, cannot be reviewed, are uninterruptable and are presented at the same pace for all students. Bates writes, "A student cannot reflect upon an idea or pursue a line of thought, without losing the thread of the programme itself. A student cannot go over the same material several times until it is understood". In developing countries, there is however, the problem of accessibility to television. In India, the vast majority of the students of Open University are not likely to have television in their homes. In such circumstances, they have to use either a community television facility or go to the nearest Study Centre for watching the television programmes.

The emergence of the new forms of electronic media like teleconferencing is of particular interest for distance learning because of their interactive capabilities. The use of electronic media components as an essential part of the multi-media approach to distance learning has always been a dream for those working in developing countries.

Teleconferencing is an interactive electronic communication among two or more locations. The punch word here is interactive and it is indeed this quality of interactivity that really distinguishes teleconferencing from the conventional one-way media like broadcasting or narrowcasting.

Teleconferencing is an already accepted form of technology in advanced countries, especially for business communications. It is only a matter of time for this technology to become popular in developing countries as necessary infrastructure like telephone links and satellite are being made available. Usually, the attraction for adopting teleconferencing is said to be the savings it results in minimizing physical travel. The time-saving involved in this process is also a significant factor. Experience of advanced countries in this process is also a significant factor. Experience of advanced countries in which teleconferencing is in wide usage shows that apart from economy and time saving, this technology improves corporate efficiency and enables participatory management.

The need for communication, staff development, and expert resources makes teleconferencing a particularly appropriate means for distant education institutions. The use of telecommunication to facilitate two-way communication among a number of students located at different venues is the basis of educational teleconferencing. The opportunity for introduction between the experts and the students, which teleconferencing so readily affords, has made it a particularly valuable learning technique.

Teleconferencing–The IGNOU and NOS Experiments

The two apex distant learning institutions in India, IGNOU and the National Open School have, during the recent years, initiated efforts to incorporate teleconferencing as a component of instructional and management methodology.

The teleconferencing experiment of IGNOU, from 4-13 October 1993, essentially a one-way video and two-way audio—was conducted to explore the feasibility of telecommunication based interaction in Distance Education. It was one of the components of the Extended Contact Programme (ECP) of PG Diploma Course of Higher Education. The two major objectives of this experiment were: to understand the potential of telecommunication in distance education and to get feed back regarding multimedia approach using talk back system.

The training programme was conducted through live lectures and discussion mode, supplemented by computer-based graphics and charts, audio-visual aid and discussion. During the discussion, the participants used STD telephone facilities for asking question in audio mode. The questions were answered using the audio and video mode.

The IGNOU Experiment was the subject of study by Sahoo (*Indian Journal of Open Learning,* Vol. 3, No. 1); and Bhatia, *Media Asia* 1995 (Vol. 22, No. 4). It was concluded : Several objectives were achieved through teleconferencing viz., clarifying doubts of participants in their studies, giving direction for different activities and reviewing the progress of activities during ECP, encouraging interaction between participants and resource persons, etc.

Bhatia suggests : Interactive narrow casting networks can be the most effective tool for HRD and training. These networks have the capability of reaching out to a very large number of selected audiences with common interests, and provide them with information and education. They have interactive capabilities, and therefore, the learning from such networks can be very high. Such networks can have very great potential for human resource development for sustainable development in the region. The establishment and utilization of such networks at the regional level need to be seriously explored.

Teleconferencing–NOS Experience

Open Schooling has evolved as one of the successful experiments in alternate schooling in the last 2-3 decades. Many have remained outside the purview of formal schooling due to socio-economic, geographical and such other reasons. Open Schools adopt distance learning methodologies which include self-learning printing materials and audio-visual cassettes; telecommunication and multimedia approach to teaching with the possibility of learning at one's pace, place and convenience has helped in catering to the individual difference of the learners.

India is one of the pioneers in Open Schooling in the world. The Government, in the Ministry of Human Resource Development, established the National Open School in 1989 for providing an opportunity for schooling to dropouts, young girls, adult women, ex-servicemen and handicapped persons. NOS has a diverse school profile having young and old learners, ranging from 14 years to as old as 89 years, living in different parts of the country. More than two hundred thousand students have so far enrolled in NOS through 800 study centres.

The NOS is confronted with a huge task of training teachers from the traditional system who needed orientation and training to perform effectively in the open learning system. The NOS conducted the training programme for 180 teachers drawn from different districts of Gujarat for 3 days from 17-19 December 1996. The mode of discussion was through satellite-based and telephone-based one-way video and two-way audio conferencing. The interactive network comprised one studio teaching centre at ISRO, Ahmedabad and six receiving centres spread over the state. The experience gained would help NOS to use this technology more effectively in a massive countrywide operational system.

US–A TALEEM Initiative

TALEEM Research Foundation, a literary, scientific and charitable society and trust, set up in May 1996 has launched a unique initiative to provide wide and easy access to learning opportunities via satellite television and information technology. The US–University in the Sky, is proposed to be set up under the Private Universities (Establishment and Regulation), Bill 1995. The US will launch multimedia and computer technology based courseware for skill upgradation, capacity building and training programmes customised to suit industry and institutional needs.

Virtual Reality, Virtual Society and Virtual University

The idea of Virtual Reality–according to the *Virtual Reality Primer,* has been around for over 25 years, but suddenly the term is cropping up all over the place–in trendy magazines, on cable news networks, and, now in academic journals as well. Interestingly the dictionary defines virtual, "as existing or resulting in essence or effect though not in actual fact, form or name." and reality is "the quality or state of being actual or true". As technology, Virtual Reality is being defined as, an interactive computer system so fast and intuitive that the computer disappears from the mind of the user, leaving the computer-generated environment as the reality.

Virtual Reality is a creative communications medium for

everyone. It will influence how we design things and run our businesses, how we teach our children and treat our illnesses, as well as how we spend our leisure time. Virtual reality environments or scenarios can be pre-defined and focused so that a user can gain specific skills or insights as if he or she were actually on the job.

In one of his papers in *Media Asia,* Dahlan says : The excitement and enthusiasm to get plunged into the information highway, for instance has propelled a part of our own societies (A small segment, to be sure, but the most influential segment nonetheless) to live in two societies at the same time; the real society in our own country and virtual society. To live simultaneously in both real, physical space and in cyberspace. We may not know or be able to evaluate which is real and which is virtual.

Virtual Reality attempts to simulate real life experiences by projecting digitally created scenes on the human brain by various means. If realized to the fullest extent its capabilities would be mind blowing and its implications staggering. From Virtual Reality to Virtual Classroom and Virtual University is natural extension of application of communication technology to education.

Interestingly the distance education theorists during the recent years have attempted to classify the use of teaching aids and technologies into generations. A general classification is as under :

1. Generation I–Mailable materials : including print material and audio and video cassettes.
2. Generation II–Education through Air : Broadcasts, Telecasts, Talkback TV.
3. Generation III–Computer based teletechnologies : CD-ROM, Multimedia and Internet.
4. Generation IV–Technologies of Immediate Future : Virtual Classroom

Desmond Keegan from the University of Dublin, in an interesting paper, recapitulates his observations about teaching and learning in a European Virtual Classroom. Keegan defines Virtual Classroom as :

A Virtual Classroom comprises an electronic classroom from which the class is taught, a network of specially equipped electronic classrooms at which the students are present, and the satellite, microwave or cable linkages between them. For teaching purposes, virtual classrooms can either be two-way video with two-way audio, often called videoconferencing or more acurately video teaching or one-way video and two-way audio.

The paper details the above mode of education employed for the, 'Certificate in Safety and Health at Work', by the University College of Dublin during 1993-94. This was one-year, part-time study programme of about 150 hours. Lecturers would deliver lectures in a broadcast standard educational television studio at the university. The sound and vision output was sent by a permanently installed microwave link to the national television transmission centre. From here it was retransmitted by the EBU uplink to Eutelstate and then downlink to the satellite dishes at colleges and centres and other receiver sites. The didactic structure of the course was paid great attention. An observation from the conclusion is a pointer towards the expanding concept of virtual classroom and its application in future :

> Transnational teaching face-to-face at a distance is now a reality and successful systems have been reported in the literature. There is, for instance, cable from Dublin to Stockholm, so it is entirely feasible to set up a two-way video, two-way audio virtual classroom between the two cities. For the first time a student fidgeting in Stockholm could disturb the teacher and the rest of the class in Dublin or a student in Dublin could interact with the teacher in Stockholm. Eutelsat II F4 has a foot print that covers the whole of Europe and viewers from Netherlands, France and Italy participated live by telephone in the 1993-94 academic year course. Students can now participate in an international classroom in which students from all over Europe are classmates.

Another instance of the working of Virtual Classroom at Monash University's Gippsland campus in Victoria, Australia has been reported in his paper, Virtual Classrooms : Real, Accessible, User-friendly, by Neil Hanley in *Media Asia* (Vol. 22, No. 4, 1995).

Hanley discusses about Net Face (Network Interface), an evolving piece of computer software which provides on-campus, distance education and mixed mode students in some semester-subjects with easy access to a variety of online services. Hanley had a more ambitious aim–to converge the on and off-campus students and mixed mode populations in a virtual classroom and believed Netface could evolve to help implement the concept of a virtual university based on group and collaborative, as well as individual learning.

Netface is still in a prototype stage of implementation, He adds : When all students can access the graphic power of Internet, when they can retrieve (cut and paste) information electronically, when they can search the information universe for what they wish to know, then the virtual classroom will be a reality.

Asynchronous Learning Networks

Dr. Burks Oakley, Professor at the University of Illinois has reported a more comprehensive pilot experiment in creating virtual classroom, as well as 'Virtual Learning Community', at the University of Illinois.

The pilot essentially involves implementation of several innovative uses of computers and computer networks to develop a new pedagogy for delivery of university engineering courses. Since the approach involves asynchronous access to networked learning materials, the initiative has been called as "Asynchronous Learning Networks".

The faculty at the university realised that with the increased access to networked computers on the campus, the students did not have to depend on "stand alone" courseware; software programme was expanded to include electronic submission and grading of all home work assignments and quizzes over the Internet computer network.

Dr. Oakley states : It seems that this approach is improving student learning. The network-based homework gives students immediate feed back. Rather than have to wait several days for their homework to be graded, students learn the results immediately. If students do not get the solution completely correct on the first attempt, then they are motivated to find their mistakes and submit the new answers. Another pedagogical feature is that students have a home work deadline every five days (on average). This encourages them to study the course material on regular basis.

In the, *Vision of the Future* Dr. Oakley concludes : Asynchronous, network-based, interactive learning environment should find more uses in the future, as more students have access to computer network where they live and study. This technology may permit us to offer self-paced courses to all students; individual students would be able to use the network to communicate with other students at a similar stage in the course. Our network-based approach should be useful in the delivery of continuing education courses to the engineers. With the explosive growth of the World Wide Web, we can expect to see many of the key features of our interactive learning environment be implemented on the web, where they will be more universally accessible.

The above three instances from Ireland, Australia and USA do give an overview of how concepts of Virtual Classroom and Virtual University are beginning to emerge as distinct possibilities as a result of innovative use of new information and communication technologies in education.

Virtual Online University (VOU)

In the USA the idea of Virtual Classroom and Virtual University has already taken off in the right earnest. David Hawkridge mentions the existence of Virtual Online University in the USA which was established in April 1994.

The Corporate Mission Statement of Virtual Online University is:

- to provide low-cost, high-quality education;
- to offer distance education using interactive, interdisciplinary methods outside of traditional learning paths;
- to assist traditional and non-traditional learners in furthering their educational objectives;
- to form working relationships with business and industry to provide students with valuable hand-on experience to supplement a liberal arts education; and
- to conduct research and provide a practical forum for investigation of online environments and applications in distance education, telecommunications and electronic delivery systems.

According to Dr. William Painter, Executive Director, VOU will target two primary audience : first, current college and university students in traditional education paths who wish to broaden their opportunities with online education; and second is the non-traditional, learning-disenfranchised person, that is, individuals who have limited access to traditional education due to financial restrictions, physical challenges, being part of at-risk population, or with responsibilities which preclude pursuing educational paths.

Apart from the above VOU, dozens of school districts in the USA are experimenting with intranets on local networks (LANs) as carriers of interactive multimedia formal educational content. The rapid spread of Internet has given a new edge to the possibility of

Virtual Classroom and Virtual University in the Asian region. Even in a country like India, which is quite down the line with regard to access to telecommunications and computers, corporate offices, commercial and media establishments are up beat about 'Virtual Office'. Although the scenario presently is uncertain but the following excerpt from the *Dataquest* (July 1994) does indicate the trend, however tentatively :

Outdated telecom facilities, high cost of connection, a waiting list that never ends, low technological awareness, bureaucratic interference . . . can any kind of revolution ever take place in India? and adds, "A sane person would think twice before betting that India would be a part of the information superhighway. But, five years down the line chances are the country would have networks similar to the best of the world.

While five years may be too early a period, it is certain that use of Internet for education in India has made an earnest beginning. The nationwide computer network of the Planning Commission, NICNET (National Informatics Centre) is linked through satellite dishes around the country. It is also connected to Internet and other international networks and databases. The Education and Research Network (ERNET) connects the Indian Institutes of Technology and other prestigious science and research institutes. The facsimile facilities have now become common and electronic mail is becoming available even in some district towns, apart from the state headquarters. That facilitates progress towards creating a Virtual University.

Creating a Virtual University in Asia–A Vision

In the Asian region dissemination and sharing of knowledge within the structured education system are still basically class-room oriented, paper-intensive processes. Those institutions which take advantage of advances in technology, especially those linking their schools to powerful networks, rich data bases and instructional equipment, will be able to provide quality educational opportunities to greater numbers and perhaps ultimately in a cost effective manner.

Let us look at the ground realities in the realm of access to tele-communication in Selected Asian Economies in the table below (excerpted from Heather E. Hudson's paper–Restructuring the Tele-communication Sector in Developing Countries–Lessons from the Asian Experience, *Media Asia* (Vol. 24, No. 3).

Two important situations emerge : The contrasts between

Access to Telecommunications in Selected Asian Economies

	Tel Lines /100	Cellular /100	Pay Phone /100	TV Sets /100	Computer /100
The Giants					
China	2.9	0.13	0.23	23.1	0.15
India	1.31	0.01	0.16	5.5	0.10
The Four Tigers					
Hong Kong	54.0	7.40	0.77	35.9	11.32
Rep. of Korea	39.7	2.16	6.87	32.4	11.25
Singapore	47.3	8.36	10.85	38.0	15.26
Taiwan	40.0	2.75	5.71	31.5	8.09
Emerging Economies					
Southeast Asia					
Indonesia	1.3	0.04	0.41	14.7	0.28
Malaysia	14.7	2.93	3.28	23.1	3.28
Philippines	1.7	0.30	0.09	12.1	0.52
Thailand	4.7	1.10	0.72	24.7	1.16

Japan, Singapore and Hongkong, with among the world's highest densities of telephone lines and cellular phones and countries such as Indonesia and Philippines is quite apparent. China and India account for just 12 per cent of the wealth and telephone lines, but have two-thirds of the region's population.

In many Asian countries, television is often more accessible than telephone services. China and the Philippines have more than seven times as many television sets as telephone lines, while Indonesia has 11 times as many. The pace of proliferation of television sets is faster. With greater use of modes like teleconferencing and video cassettes, even for the school level education, Asia is crertainly ready for more tele-education.

Creating a Virtual University in the Asian context would continue to be a vision for a large number of Asian countries. Technologies such as audio cassette players, television and VCR, even though familiar to the students, constitute essentially leisure and entertainment media and till such time as sustained efforts are made to reorient these technologies for instructional purposes, these may not be adequately used as such.

ENVIRONMENTAL REPORTING FOR SUSTAINABLE DEVELOPMENT

Over the past 25 years and more, in the face of emerging challenges, activism of social action groups and non-governmental organisations and threat of globalisation to our natural and genetic resources, Environmental Reporting for Sustainable Development, has assumed added urgency. This is more so in the wake of shifting paradigms of development and organised ecology movements against construction of dams and large river valley projects, which, not so long ago, were considered as 'temples of modern India.' The new awareness crystallizing into a number of initiatives at the national and international level, as well as judicial activism, has given Environmental Reporting a measure of respectability, so far considered of peripheral importance.

A refreshing insight developed during the recent years is in relation to environment, development and gender issues. Environment is no longer a strait-jacket subject. Our understanding of the relationship between environment and development has undergone a profound change since the 1972 Stockholm UN Conference on Environment. At the end of the 1960s, it appeared to all, but the most enlightened that we could have one or the other, and if we wanted to have development then the price to pay would be a loss in environmental quality. This view has largely been changed as we have come to realize that the two-environment and development are interdependent. Without conservation one cannot have development and without development no conservation. Since the 1972 Stockholm UN Conference terms such as *eco-development, sustainable development* and *environmentally sound development,* have come into widespread use. It has now been largely understood, and accepted that long term development, can only be achieved through sound environmental management.

Concern for environmental degradation has been enlarged to include issues such as women, their land rights and health. The Dag Hammarskjold Foundation, underlined the importance of the issue by devoting its 1992 number of *Development Dialogue,* to Women, Ecology and Health-Rebuilding Connections. The noted environmenta-list, Vandana Shiva in the Introduction to the issue says :

> Across the world, women are rebuilding connections with nature, and renewing the insight that what we do to nature, we do to ourselves. There is no insular divide between the environment and our bodies. Environmental hazards are also health hazards in food systems. Pesticides do not merely pollute fields, they end up polluting our bodies. Destruction of biodiversity does not merely impoverish nature, it impoverishes

tribal and peasant societies. These links exist in the real world even though they have been denied by fragmented and divided world views. Beginning with women's experiences, analysis and actions we will rebuild the connections between ecology and health, for a more holistic approach to the contemporary crisis of survival.

Environmental Reporting

Environmental protection has become contemporary crisis of survival, no less. It would therefore be essential to go into what Environmental Reporting is all about. To define Environmental Reporting just by issues may be too narrow, on the other hand a broader definition would mean that since environment involves the active relationship between people and nature at large, there is very little which may be excluded from Environmental Reporting. In this context it would be pertinent to refer to the noteworthy initiative taken by ESCAP in bringing out a handbook for journalists, Reporting on the Environment. Apart from elaborating the concept of Environmental Reporting the handbook gives valuable information about environmental issues such as acid rain, greenhouse effect, ozone and the atmospheric ozone layer, etc. A glossary of technical and scientific terms alongwith notes and references, adds value to the publication.

The handbook defines Environmental Reporting as including coverage of health economics, politics, commerce, development, physical resources and a wide variety of scientific research. In short, people's interactions with the living and non-living components of their environment, which alter the environment by using, redistributing or disturbing its resources, constitute a topic for environmental reporting. International problems worth covering include the ever present growth of human populations, poverty and lack of food to feed increasing numbers of people and the interrelated problems of environmental destruction, deforestation, desertification, soil erosion and habitat destruction. International positive activities include the concerted efforts led by the United Nations to protect the ozone layer, international treaties to protect endangered species and cooperative efforts among nations to help to transfer appropriate technologies. Nationally, the problems of increasing urbanization and the accompanying air, water and noise pollution are also part of the environmental beat. Positive national activities would include sustainable development planning and projects that take into account preservation of natural resources.

In China, there is a great interest in the specialized Environmental Reporting. Apart from a large number of environmental journals and publications at the provincial, municipal and regional levels, China Environmental News, a national newspaper has a circulation of 600,000 copies. In the United States, many newspapers created environment as a special beat when environmental concerns were just emerging in 1970s.

In India media played a notable part in carrying forward the message of the *Chipko* movement for forest protection and regeneration which began on 24 April 1973 in the hills of Uttar Pradesh-Mandal near Gopeshwar. Today *Chipko* has become an important term in the lexicon of environmentalists the world over. The *Chipko* movement, which received sustained media attention in the country, is probably the world's most well known grass roots eco-development movement. *Chipko* activities have attracted the attention of environmentalists, Gandhians, various persons interested in the process of social change and the media persons throughout the world. The non-violent, action oriented *Chipko* movement has greatly helped unite the people and focus attention on the mismanagement of forest resources. Its Gandhian character has brought it considerable sympathy. One line of thought is that the main objective of forest management should be soil and water conservation : forests do not bear timber, resin and foreign exchange but soil, water and pure air. The self-sufficiency of the hill people in food, clothing and shelter is important but is secondary to the major ecological objectives. This cannot be achieved unless all commercial green felling is stopped and no new contracts entered into with industrialists to supply raw material. Another philosophy behind the *Chipko* movement is the belief that no conservation is possible without the cooperation of the people. Eco-development strategies will inspire the poor when they are based not just on ecology but also on the needs of the people.

Crusader for a cleaner environment, human rights and justice, and a development journalist Bharat Dogra whose several publications on environment including, *Encyclopaedia of Environment,* have greatly contributed to the cause of environmental awareness, gives a graphic description of the involvement of ordinary villagers with the movement– high voltage emotions, villagers with axes and saws on one side including women and children; and labour-contractor, the provincial armed constabulary force men, 100 strong and forestry department officials on the other. The essay-

Environment: Grassroots Movement, written for *India at 50, Bliss of Hope and Burden of Reality,* an immensely readable book, shows that the environmental movements are bound to succeed because they essentially address grassroots concerns of the peasants, forest dwellers and fisher folk. The *Chipko* movement which began with the stopping of felling of trees and sawing jungles has successfully involved people in greater issues of ecology and environment.

Environmental Movements

The *Chipko* movement later on inspired the *Appiko* movement in Uttara Kannada, which forms the part of western Ghats and is known as the forest district. Waking up to the destruction of forests, the local people especially the youth were motivated into direct action. A group of youth in Balegadde village, protesting against moves to establish teak plantations, wrote to forest officials asking them to stop clearing the natural forest. But this appeal was ignored. Then the villagers decided to launch the *Chipko* movement, popularised by environmentalist Sunder Lal Bahuguna in the Himalayas. Bahuguna was invited and people took an oath to protect the trees by embracing them–This came to be known as *Appiko* in Kannada. In September 1983, when the axemen came for felling to the Kalase forests, people embraced the trees and thus the *Appiko* movement was launched. This simple non-violent action became popular statewide and people joined this movement spontaneously in many forest areas of the Western Ghats.

The movement spread to Coorg, Dakshina Kannada and Shimoga districts. In Coorg, people stopped the felling of trees in the Cauvery catchment areas, for which the Government had given felling permits to plywood units. The people camped in the forests for months to save the natural wealth. The success of the *Appiko* movement is because of the constant interaction with villagers by encouraging their participation in all programmes. Through people's participation, *Appiko* is trying to evolve a sustainable development strategy that conserves and improves natural resources such as forests.

The Narmada Bachao Andolan (NAB) is another well known of the environmental movements. The importance of the movement can be assessed from the comments made by Pravin Sheth in his book, *Environmentalism–Politics, Ecology and Development* in the section-Narmada Project : Major states of western India provide a most important case to study in terms of maturation of environmental movement and dynamics related to politics of development. No major project in India has brought the ecodevelopment problems to such a high level of informed debate, emotive rhetoric, political mobilisation, and grass roots activism as this project has. The controversy around this project has engulfed horizontal and vertical levels of government and authority and helped forge linkages between NGOs operating at grass roots to national and international levels. It has also contributed to the political discourse on development alternatives–in fact, becoming the ideo-political nursery for greening the politics of India. Whereas the project was considered absolutely indispensable for the water-starved northern and Saurashtra regions of the state, its critics termed it as "the world's worst man-made ecological disaster."

There has been growing resentment and anger against different aspects of development on the pattern of dominant paradigm, which gives primacy to infrastructure on western model. Well known Gandhian economist L. C. Jain, political activist Srilata Swaminadham, eminent environmentalist Sunder Lal Bahuguna and activist Vandana Shiva, through their writings, interaction with grassroot workers, tribals and peasant societies have brought forth refreshing and rich body of knowledge. An environmental journalist will have to take note of the alternative models of development, different from those advocated by IMF-WB protagonists. These social thinkers have been raising issues : inspite of billions of dollars spent over the last 40 years or more, why the number of poor is rising, and why environmental and ecological damage is becoming irreversible.

The following excerpt from *Save Himalayas* by Sunder Lal Bahuguna subsums the above thinking on environmental protection: Environment should not be seen in isolation. Similarly, conservation is not possible till we adopt a life style which had been developed for generations and reflectes the wisdom of the people. Wisdom is the collection of experiences, which common people have gained for generations. New generations have made new additions to it. This traditional wisdom has been ruthlessly suppressed by the modernity which is based upon a philosophy of development with economic growth. In the religion of economic growth, market is temple, the technocrats and experts are the priests and dollar the God. This has taken humankind on the brink of disaster. Remote mountain areas are the frontiers, which have till now protected our natural heritage, specially mountains, glaciers, forests, rivers, meadows, etc. Modernity looks into these as a commodity and these to accelerate the pace of economic growth. This is exacting the pound of flesh.

Of late, thinking among Third World development NGOs and workers has been highly critical of the purpose and ultimate utility of international and inter-government conferences, which are a highly hyped affair when these are conceptualised, and later on found to have been short in respect of their objectives, fulfilment of resolutions and implementation of decisions taken in these fora. The 1992 Earth Summit in Rio which had received top media billing at the international level and was heralded as the 'last chance to save earth', the implementation performance of the summit and record in terms of follow up action has been nothing to write home about. In contrast, the Global Meet in Tilonia, Rajasthan, 10-14 March 1997 which was attended by volunteers and workers from non-government agencies working at grass-root level from 38 countries, could work out alternative strategies for healthy and sustainable planet.

Tilonia Declaration

The participants at the Tilonia Global Congress held the view that global conferences should not be reduced to level of "expensive talkshops" that have no effect on their lives. The representatives at the meet who were mostly volunteers drawn from villages were actually aware of the problems of the poor, the tribals and the village people in the wake of rapid deforestation, quarrying and mining in the hill areas in several countries. The underlying idea enunciated at the Congress was that initiative be undertaken to counteract the current global trends of market economy and globalisation in which the modern develop-mental paradigm promoted a life style and thinking detrimental to the Third World countries.

Some of the important recommendations of the Tilonia Declaration are :

Experience has shown that global conferences have done little to set an example of austerity and simplicity. They have had little tangible impact on the lives of the poor. Such conferences must be avoided.

No sustainable solution to improve, preserve, protect and conserve the environment is possible without the direct involvement of communities in the planning and implementation process. Space must be created for the oppressed communities to participate in decision making. Any move to by pass them leads to wastage of funds.

The learning and unlearning process of governments, donors, planners and experts begins with the awareness and recognition that communities themselves have the capacity to identify and solve their problems. Consultants are required only marginally. It is foolish to think that environmental problems could be solved without respecting the knowledge, skills and wisdom of the people themselves.

Campaigns and non-violent struggles are the only legitimate and democratic means left of pressurising governments and donor agencies to stop the pollution and destruction of the environment. Numerous examples were shared at the congress where status quo governments have been forced to change policies and actions.

That every citizen has a right to information on public expenditures incurred in their own villages and slums. Governmental and non-governmental agencies must be accountable and transparent to the communities they affect.

Where industries are being established, local communities must have the right to monitor them for environmental safety. That the choice of technology must not over exploit or degrade natural resources, displace jobs, marginalise indigenous people or increase dependencies on funds and expertise from outside. Resources must be redirected from polluting dangerous and capital intensive technologies towards the development of ecologically sustainable solutions.

The causes of rural urban migration, the process of urbanisation and growing consumerism must be addressed and we need to ensure that there is broad based participation in decision-making, planning and development of more people-centred and sustainable human settlements.

That there is increasing evidence of spiritual poverty on environmental issues which need to be addressed by all stakeholders. We endorse the Code of Environmental Ethics and Conduct.

We commit ourselves to extending demonstrative and practical examples of effective grassroots environmental actions such as the eco-volunteer concept which promotes the recognition and support of local activists in their environmental struggles.

That the widespread dissemination of the concept of eco-volunteer as an innovative, cost effective way of tackling environmental problems at the grassroot level is an immediate necessity.

The state of biodiversity and healing of ecosystems must be recognised as critical indicator in any alternative development model and given top priority. The practice of analog forestry of the enhancement of biodiversity must be promoted.

It must be recognised that all actions have a social, cultural and political dimension. These must be included in educational curriculum.

Judicial Activism

Public interest litigation, launched by social activists to protect green belt in towns from enchroachment at the hands of Municipal Corporations and builders, seeking the intervention of the courts on different issues–right from protecting the Taj to introduction of lead-free petrol, and other environmental related problems. The fact of the matter is that the courts of the country have come to the rescue of environmentalists where the local administration or the government failed to take effective measures to enforce legal provisions. Reporting of such cases has led to more active interest in protecting the environment. Likewise the ineffective working of the Boards for Prevention and Control of Water Pollution, lack of political will to enforce anti-pollution laws, loop-holes in the anti-pollution legislation have helped create a body of public opinion alert to meet the situation.

Apart from pioneers in environmental movements, social activists, environmental journalists, writers and researchers, several lawyers and judges have contributed magnificiently in this battle against degradation of the environment. The Delhi-based public interest lawyer, M.C. Mehta, who has successfully fought and won 40 landmark environmental cases in the Supreme Court since 1984, is one such activist. In a profile of the grassroot crusader, Mehta, who won the 1996 Goldman Environmental Award, the *Times of India* in its issue of 26 May 1996 recapitulates his major success stories. Mehta believes that a strong people's movement is essential in the country, political parties must be forced to give priority to environmental causes and an independent national environmental commission, able to withstand pressure groups, must be set up.

Environmental journalists have their hands full. They will not only have to continue taking sustained and informed look at whatever is happening in the vastly expanding and complex field of environment; but will also have to maintain a Crusader's interest as to how the government and number of public funded bodies are doing or not doing. There is no alternative to an informed public opinion about issue of environmental protection which only a vigilant and watchful media can help create.

WOMEN EMPOWERMENT–ROLE OF MEDIA : BEIJING AND BEYOND

The evolving movement of women empowerment, together with the recognition of women's rights as human rights, has been one of the dogged and determined public campaigns, and substanially successful, during the last quarter of the twentieth century. This movement has been one of the important social movements, environmental protection and human rights being the other important ones; which calls for worldwide strategy of equality, sustainable growth and justice. The movement has been a long time in the making and can be traced to the growth of women's movements globally during the UN Decade for Women from 1975-85. Since then the women have been constantly questioning as to why women's rights and lives have been deemed secondary to the lives of men. The role of women, as a definite issue, entered international debate at the first UN Conference on Population and Development, held in Bucharest (Romania) in 1974. Since then the dimensions of the problem have become more complex. Other conferences followed, with the contribution of rural women to agrarian economies acknowledged by the World Conference on Agrarian Reforms and Rural Development (1979). The African famines of the time raised the question whether they might have been caused, too, by the neglect of women's role as primary food producers. The Conference on Human Rights held in Vienna (1993) agreed that women's rights are human rights.

The UN Decade for Women considerably succeeded in generating concern on women's issues both at the national and international levels. Besides emergence of autonomous groups and developmental organisations, gradually realizing the need of women's participation in the developmental activities the decade promoted academic pursuits, particularly research on hitherto ignored aspects of women's lives. The International Conference on Population and Development (ICPD) (1994) held in Cairo, Egypt, was geared to ensure that population and development policies and endeavours emphasized women's equality. Education, employment opportunities and health services including reproductive health care and family planning were treated as crucial components of women's advancement. ICPD stressed gender equality, ending violence against women and eliminating administrative, cultural, social and economic obstacles to the realization of women's rights and economic independence.

The position adopted at the Social Summit in Copenhagen (1995) made the empowerment of women part of the basic commitment of some 120 heads of State or Government. Indeed, the summit at Copenhagen explicitly merged :

- Women's rights into the broader convenant on human rights; and
- the central role of women into social development and thus into their own empowerment.

The International Year of Women (1975), the Decade of Women and the strategies adopted during the three Women's World Conferences stimulated the establishment of national machineries for strengthening the empowerment of women. This brought about a higher degree of awareness of the overt and covert biases against women in all fields of activity.

Empowerment Defined

The term 'empowerment' has been variously understood as a process, a movement, collective action, etc. In the area of land rights of women, where the term carries economic connotations, Bina Agarwal, in her book–*A Field of One's Own, Gender and Land Rights in South Asia,* defines empowerment 'as a process that enhances the ability of disadvantaged (powerless) individuals or groups to challenge and change (in their favour) existing power relationship that places them in subordinate economic, social and political positions'. Empowerment can manifest itself in acts of individual resistance as well as in group mobilization. Entitling women with land would, on the one hand, empower them economically, and, on the other hand, strengthen their ability to challenge social and political gender inequalities. Land rights would enhance women's freedom to achieve (or capability to function) in non-economic spheres as well.

Indeed the term empowerment has been equated by some with collective action. However, as a definition of empowerment thus would appear too narrow. In a limited sense, collective action may itself empower women by enhancing their self-confidence and their ability and willingness to challenge oppression, but in a larger sense it is a means to empowerment, wherein empowerment lies not only in the process of challenging gender inequality but in eliminating it.

In this sense, collective action is one channel for change. What is important, nevertheless, is that it is often the most effective one.

The noted sociologist Yoginder Singh says : Empowerment of women conceptually forms a distinct category by itself. It implies processes by which women's power of self organisation is promoted and reinforced, they develop the capacity for self-reliance outcrossing the relationship of subordination on account of gender, social and economic status and the role in the family and society. It encompasses their ability to make choices, control resources and enjoy participatory relationship within family and community. To achieve these objectives empowerment of women also implies their ability to consolidate social movements to remove obstacles in the path, be it from state and administration, family and kinship, workplace and its organisation, institutions of production of culture and knowledge or arenas of politics and power.

For the overall development of society, women's empowerment and their full participation on the basis of equality in all spheres of society including participation in the decision-making process and access to power, are fundamental for the achievement of equality, development and peace in this world. The empowerment and advancement of women, including the right to freedom of thought, conscience, religion and belief, contribute to the moral, ethical, spiritual and intellectual needs of women as well as men, individually or in the community, thereby guaranteeing them the possibility of realizing their full potential in society and shaping their lives in accordance with their own aspirations.

Women's Empowerment and Media : Fourth UN Conference on Women–Beijing, 1995

Recognising the role of media in Women's Empowerment, the Fourth World Conference on Women held at Beijing in September 1995 devoted a separate section to 'Women and Media' in the final Declaration and Platform for Action adopted at the conclusion of the Conference. It clearly defined the significant role media needs to play in accelerating the process of advancement of women, the document noted.

During the past decades, advances in information technology have facilitated global communication network that transcends national boundaries and has an impact on public policy, private attitudes and behaviour, especially of children and young adults. Everywhere,the potential exists for the media to make a far greater contribution to the advancement of women.

More women are involved in careers in the communications sector, but few have attained positions at the decision-making level or serve on governing boards and bodies that influence media policy. The lack of gender sensitivity in the media is evidenced by the failure to eliminate the gender based stereotyping that can be found in public and private, local, national and international media organizations.

The continued projection of negative and degrading images of women in media communications–electronic, print, visual and audio-must be changed. Print and electronic media in most countries do not provide a balanced picture of women's diverse lives and contributions to society in a changing world. In addition, violent and degrading or pornographic media products are also negatively affecting women and their participation in society. Programming that reinforces women's traditional roles can be equally limiting. The world-wide trend towards consumerism has created a climate in which advertisements and commercial messages often portray women primarily as consumers and target girls and women of all ages inappropriately.

The Declaration further adds : Women should be empowered by enhancing their skills, knowledge and access to information technology. This will strengthen their ability to combat negative portrayal of women internationally and to challenge instances of abuse of the power of an increasingly important industry. Self-regulatory mechanisms for the media need to be created and strengthened and approaches developed to eliminate gender biased programming. Most women especially in developing countries, are not able to access effectively the expanding electronic information highways and, therefore, cannot establish networks that will provide them with alternative sources of information. Women, therefore, need to be involved in decision making, regarding the development of the new technologies in order to participate fully in their growth and impact.

In addressing the issue of the mobilization of the media, governments and other actors should promote an active and visible policy of mainstreaming a gender perspective in policies and programmes.

Recommendations of the Conference–Access of Women to the Media

The Beijing Declaration and Platform for Action enunciated specific steps which governments of participating countries need to undertake. To ensure increased participation and access of women to the media, various actions to be taken by the governments, as suggested at Beijing, included :

(a) Support women's education, training and employment to promote and ensure women's equal access to all areas and levels of the media,
(b) Support research into all aspects of women and the media so as to define areas needing attention and action and review existing media policies with a view to integrating a gender perspective,
(c) Promote women's full and equal participation in the media including management, programming, education, training and research,
(d) Aim at gender balance in the appointment of women and men to all advisory, management, regulatory or monitoring bodies, including those connected to the private and state or public media,
(e) Encourage, to the extent consistent with freedom of expression, these bodies to increase the number of programmes for and by women to see to it that women's needs and concerns are properly addressed,
(f) Encourage and recognize women's media networks including electronic networks and other new technologies of communication, as a means for the dissemination of information, and the exchange of views, include at the international level, and support women's groups active in all media work and systems of communications to that end,
(g) Encourage and provide the means or incentives for the creative use of programmes in the national media for the dissemination of information and the exchange of views, include at the international level, and support women's groups active in all media work and systems of communications to that end,
(h) Encourage and provide the means or incentives for the creative use of programmes in the national media for the dissemination of information of various cultural forms of indigenous people and the development of social and educational issues in this regard within the framework of national law, and
(i) Guarantee the freedom of the media and its subsequent protection within the framework of national law and encourage, consistent with freedom of expression, the positive involvement of the media in development and social issues.

Portrayal of Women in Media

For promoting a balanced and non-stereotyped portrayal of women in the media, the Declaration suggested the following points of actions to be taken by the governments :

(a) Promote research and implementation of strategy of information, education and communication aimed at promoting a balanced portrayal of women and girls and their multiple roles,
(b) Encourage gender sensitive training for media professionals, including media owners and managers, to encourage the creation and use of non-stereotyped, balanced and diverse images of women in the media,
(c) Encourage the media to refrain from presenting women as inferior beings and exploiting them as sexual objects and commodities, rather than presenting them as creative human beings, key actors and contributors to and beneficiaries of the process of development,
(d) Promote the concept that the sexist stereotypes displayed in the media are gender discriminatory, degrading in nature and offensive, and
(e) Take effective measures or institute such measures, including appropriate legislation against pornography and the projection of violence against women and children in the media.

Projection of Women in Media

The National Commission on Self-Employed Women and Women in the Informal Sector in its report–*Shramshakti* studied the projection of women in media and came up with several useful recommendations. The core recommendations about Communication Network Systems in a capsule form are as under :

- The media must project the working women in the unorganized sector as workers and not merely as performing the duties of wife/daughter. They being major earners they must be projected as producers and not merely consumers. Unless they are perceived as workers, their rights will not be recognized.
- It is necessary to formulate a national policy on communication clearly stating what it aims to do, how it plans to achieve its objectives, what is going to be the time period and institutional structures through which the objectives are to be realized.

- All the groups which are concerned with the production of media, and/or connected with planning and dissemination, should structurally integrate the participation of socially aware persons. This will enable the planning group to perceive the gender perspective, and neither become blind to women's issues or err too much in the opposite direction of joining the band-wagon. Media monitoring unit should try to help the media to orient themselves to the interest, concerns and development of women in the unorganized sector.
- Frequent workshops of media persons may be organised to focus generally on problems of women, and specifically on problems of women in the unorganized sector. Further such workshops should also publicise the successful experiments carried out in the media so as to help others in developing using such strategies. Such women workers should also be participants in media workshops. This would further help in looking at the working women, not as mere objects to be acted upon, but as equal partners in creatively disseminating information.
- Innovative efforts to communicate should be encouraged. If possible, some percentage of the financial allocation of the concerned ministries may be reserved for such experiments so that more talents may emerge.
- Considering the context of poverty and unemployment for large masses of the people and, therefore, a lack of access to electronic media its overuse should be discouraged. Instead, other forms of interactions such as fairs, shibirs, health camps, jathas should be promoted.
- All the media channels, both in the public sector and the private sector should take serious note of the Seventh Plan directive that *themes which have pernicious consequences of conspicuous consumerism should not be projected. In this context, advertisments displaying women as sex symbols and using them for sales promotion should be seriously dealt with.* It is not enough to control pornography through legislative enactment. It is equally necessary not to convey sex linked division of labour or women's predilection with feminine pursuit through conservative and treaditional stereotypes. More vigorous action of policy implementation is called for.
- The public sector media should make deliberate attempts to not only project the problems of women in poverty, but should monitor in such a way that conflicting role models are not

depicted, nor derogatory references to their work are made.

- To improve content and coverage, coordinated efforts for increased interaction between NGO's, women's social action groups, research organizations, institutes of mass communication, and the media personnel should be developed. The Information and Broadcasting Ministry should evolve such network to monitor the projection of women, and evolve a code of ethics with regard to the presentation of women in all types of media.

All this seems to be a tentative beginning, a drop in the ocean, when we look at some of the development indicators, literacy being the fundamental for women. The illiteracy among women is endemic, literacy being 36 per cent for women compared to 64.3 per cent for men. With the exception of Kerala where the female literacy is 86.9 per cent, in the states like Bihar, Rajasthan, Uttar Pradesh and Madhya Pradesh less than 30 per cent women are literate. Almost three fourths of the population is concentrated in this region. The UNDP Report places India at 103 out of 174 in the rank of "gender related develop-ment index." On gender empowerment measure, India ranks 93 while China ranks 29, Indonesia 61 and even India's neighbours Nepal and Bhutan score better.

The Beijing Women's Conference–billed as a "Conference of Commitment", was certainly the most significant milestone in the women's struggle for empowerment. Euphoria over, it is now being increasingly realised that it was much easier to ride the crest of a revolutionary wave than help steer change in the desired direction. The women organisations would have to build up political, institutional and organisational capacity in diverse ways–building and maintaining network contacts among diverse groups, espousing causes from environment to labour, forging partnership with the academia, and using media effectively. The struggle to be able to stay in the fore-front of the fight in the midst of competing interests and issues is by no means easy. The need to go beyond rhetoric is keenly felt even in media circles.

REFERENCES

Media Practice and National Unity

Agee, Warren, K. (1969). *Mass Media in a Free Society.* Oxford and IBH Publishing Co., New Delhi.

Dreze, Jean and Sen, Amartya (1995). *India–Economic Development and Social Opportunity.* Oxford University Press.

Dreze, Jean and Sen, Amartya (eds.) (1997). *Indian Development–Selected Regional Perspectives.* Oxford University Press.

Holaday, D. (1996) The Social Impact of Satellite Television in Indonesia. *Media Asia* 23 : 102.

Hertz, Frederick (1966). *Nationality in History and Politics.* Routledge and Kaugan Paul Ltd., London.

Kunczik, Michael (1984). *Communication and Social Change.* Friedrich-Ebert-Stiftung, Bonn.

Liu, Alan P. L. (1971). Communications and National Integration in Communist China–The Centre for Chinese Studies, University of Michigan, Oxford University Press.

MacFarlane, Andrew (ed.) (1987-88). *CAEJAC Journal.* The Commonwealth Association for Education in Journalism and Communication, University of Western Ontario, London, Canada.

Mitra, Anand (1993). *Television and Popular Culture in India.* Sage Publications, New Delhi.

Qureshi, Fazal (1993). Serving the Nation. In : *Role of Media in a National Crisis.* AMIC, Singapore, pp. 69-76.

Singh, S. Nihal (1993). Painful Splits. In : *Role of Media in a National Crisis.* AMIC, Singapore, pp. 19-26.

Literacy, Educational Technology and Distance Education

Basu, Sisir (1996). Regular Telecasts by IGNOU : Feed-back from Students. In : *Indian Journal of Open Learning* Vol. 5, No. 1, pp. 34-43.

Dahlan M. Alwi (1995). Information Technology and Society. *Media Asia* 22, No. 4.

David Hawkridge (1995). The Big Bang Theory in Distance Education. In : *Open and Distance Learning Today,* Fred Lockwood (ed.). Routledge, London.

Indira Gandhi National Open University (1995). Future of Communication Technology in Distance Education, ES (318).

Keegan Desmond (1995). Teaching and Learning by Satellite in a European Virtual Classroom. In : *Open and Distance Learning Today,* Fred Lockwood (ed.). Routledge, London.

Larijani, L. Casey (1993). *The Virtual Reality Primer.* McGraw Hill Inc., New York.

Oakley Burks II (1996). A Virtual Classroom Approach to Teaching Circuit Analysis. In : *IEEE Transactions on Education,* Vol. 39, No. 3.

Outlook (1996) Vol. II, No. 27. An other recent article in *India Today* recounts the experiences and joys of three persons pursuing their studies for Ph.D. degree at Punjab University, Chandhigarh at the age of 70 years.

Reddy, G. Ram (1988a). China's Radio and Television University—World's Largest Network of Open University in *Readings in Distance Education-4.* Indira Gandhi National Open University, New Delhi.

Reddy, G. Ram (1988b). Role of the Media in Distance Teaching in *Readings in Distance Education-4.* Indira Gandhi National Open University, New Delhi.

Gopal Saksena (1997). Director, Commonwealth Educational Media Centre for Asia (CEMCA). *Television in India, Changes and Challenges.* Vikas Publisher.

Sahoo, P. K. (1994). Teleconferencing in Distance Education : IGNOU Experiment in *Indian Journal of Open Learning,* Vol. 3, No. 1, pp. 29-32 and Bhatia, B. S. (1995). Interactive Narrowcasting for Distance Education. *Media Asia* (1995) (Vol. 22, No. 4).

UNESCO. (1982). *Many Voices, One World.* Paris.
Waniezwicz Ignacy (1972). Broadcasting for Adult Education—A guidebook to World-wide Experience. UNESCO, Paris.

Environmental Reporting for Sustainable Development

Bahuguna, Sunder Lal (1994). *Save Gangotri in Save Himalayas;* Chipko Information Centre, Silyara, Tihri Garhwal.
Dogra, Bharat (1997). Environment Grassroot Movement. In : *India at 50–Bliss of Hope and Burden of Reality,* V. N. Naryanan and Jyoti Sabharwal (eds.) Sterling Publishers, New Delhi.
ESCAP (1988). Reporting on the Environment—A Handbook for Journalists, Bangkok.
India Today (1986). Deforestation–An environmental holocast. 15 August pp. 73-87.
Jain, L. C. (1997). Strengthening the foundations. In : *Seminar,* March, No. 451.
Sebastian Sunny (1997). A Global Meet with a Difference. *The Hindu,* 27 April.
Sheth, Pravin (1997). *Environmentalism–Politics, Ecology and Development.* Rawat Publications, Jaipur.
Shiva, Vandana (1992). *Women, Ecology of Health : Rebuilding Connections in Development Dialogue* (1-2), p. 3.
Shiva, Vandana (1997). Environment : The Scorched Earth. *The Observer,* 15 August.
The Hindu (1991, 1994-97). Survey of the Environment, Madras.
Tolba, Mostafa Kamal (1987). *Sustainable Development–Constraints and Opportunities.*
Venkateswaran, Sandhya (1992). *Living on the Edge–Women, Environment and Development,* Friedrich Ebert Stiftung, New Delhi.

Women Empowerment–Role of Media; Beijing and Beyond

Agarwal, Bina (1995). *A Field of One's Own–Gender and Land Rights in South Asia.* Cambridge University Press.
Dreze, Jean and Sen Amartya (eds.) (1966). *Indian Development—Selected Regional Perspectives.* Oxford University Press.
Prasad, Nandini (1996). Media Policy and Women Issues. In : *National Media Policy,* V. S. Gupta and Rajeshwar Dyal (eds.). Concept Publishing Company, Delhi, pp. 46-52.
Sadasivam, Bharati (1996). Onward from Beijing, Challenge of Accountability. *Times of India,* 31 October.
Shiva Vandana (1992). *Women, Ecology and Health : Rebuilding Connections In Development Dialogue Journal.* pp. 3-11.
Shramshakti (1988). Report of the National Commission on Self Employed Women and Women in the Informal Sector, New Delhi.
Singh, Yogendra (1997). Social Change : Rise of Popular Culture. In : *India at 50–Bliss of Hope and Burden of Reality,* V. N. Narayanan and Jyoti Sabharwal (eds.). Sterling Publishers Private Limited, New Delhi, pp. 328-344.

8

MEDIA TRENDS AND DEVELOPMENTS— An Update

Several emerging trends and developments during recent years have significantly influenced the media landscape in the country. Some of these trends are : impact of market forces and commercialisation of media, changes being brought about as a result of technological advances, trivialisation of social and cultural milieu in the society, marginalisation and dilution of the place and importance of editor in newspaper set-up, increasing corporatisation and apprehension about *murdochisation,* rapid expansion of the language press, initiative and campaign for Right to Information. While some of these have been briefly discussed in earlier chapters, an indepth discussion about some others would be worthwhile.

Media and Market Forces

The process of liberalisation under way in the country since July 1991 has had an impact on every sphere of commercial activity, and the media industry is no exception. The rapid economic growth has been accompanied by a dramatic increase in advertising. Today billions of dollars are spent on media advertising in Asia, all aimed at enticing the burgeoning middle class, thus encouraging the forces of commercialism and consumerism. The gradual build up of multinational business activity has resulted in the trasnformation of consumer behaviour as the region shifts from traditional to mass market and advertising is positioned at the centre of this change.

In one of the recent publications, *Market-driven Journalism– Let the Citizen Beware ?,* John H. McManus, an Assistant Professor in the Communication Department at Santa Clara University, and a former newspaper reporter, points out how the formerly revered practice of news reporting from public interest has been usurped by MBA/Corporate driven view that the news is a product and the reader or reviewer a customer. The media firms deploy resources, such as–reporters or photographers, to maximise return to stock holders and how such practices affect journalistic quality. Given the nature of American culture and its market-based economy, commercial media firms must compete for necessary resources in four markets–one each for investors, sources of news, advertisers and news consumers. A prominent news executive gave to the local broadcasters the advice : A marketing approach demands that we treat a newscast as a consumer durable good, a commodity that a viewer "purchases" by 'spending' time watching it. We must understand that when a viewer watches a specific television programme, the viewer really is spending a precious resource : time. And time for many people is spent as carefully as money.

Having advertising rather than consumers as primary source of income, both for newspapers and television, means that the way to increase profits is to produce a product that has a minimal threshold appeal to the maximum number of 'demographically desirable consumers' in the signal or circulation area. The economics of newspapers and television favour breadth of appeal over depth.

One of the reasons given as to why local television news falls short of journalistic standards was : stations are paid for the size of the audience of potential customers they deliver to advertisers, not on how much viewers learn about current issues and events. So, the decision to seek the largest number of viewers becomes a commercial, not a journalistic one. The public interest is subordinated to the short term interest of advertisers and investors.

At the time of its inception, and for quite some years thereafter, the mandate of Doordarshan was to telecast development programmes. Vikram Sarabhai, who in the formative years of television growth in the country, impressed upon Indira Gandhi the development potential of TV, in a paper presented at an international conference in 1969 said: National programme which would provide television to about 80 per cent of India's population during the next ten years would be of great significance to national integration, for implementing schemes of social and economic development, and for the stimulation and promotion of electronic industry. It is of particular significance to the large population living in isolated communities. However, those were still the early days of idealism, which did not last long.

Emerging Forces of Commercialism and Consumerism

The Working Group on Software Planning for Doordarshan, stressed as early as 1985, the need to prevent the appropriation of television by the emerging forces of commercialism and consumerism. While these forces slowly tightened their grip on Doordarshan, liberation came for the Indian T.V. viewers via foreign satellites. The Gulf War in February 1991 ushered in CNN. Subsequently new cable networks sprang up; Satellite Television for the Asian Region (STAR) began operation out of Hong Kong.

India is increasingly becoming the playing field of world's biggest media players for their share of our unregulated 15 million homes Cable TV market run by 60,000 operators. This market is bigger than that of any market in Europe, barring Russia and Germany. Not surprisingly, by mid 1993 India and the market it represented had caught the attention of the world's best known media marauder, Rupert Murdoch. He bought controlling shares in STAR in July 1993 and later, in December that year, 49 per cent of the equity in Zee TV.

Big money and bigger target audience-advertisements today seem more capable of swaying the masses than ever before. The Wills World Cup is remembered as much for its nail biting finals, India's incredible debacle at Eden Gardens and Sri Lanka's spectacular victory–as much as for its commercial extravaganza and glitzy advertisements.

The entertaining battle between the Coke and Pepsi almost became bigger than the tussle for the World Cup proving that advertising in India has finally gone beyond boom time. Big money, big advertisements and a bigger target audience than ever before has the ad world on an all time high. A surfeit of commercials, press campaigns and hoardings combine to form an effective visual onslaught on the senses. We end up becoming willing victims of media's most enticing side–an aggressive, glamorous and strategy.

In a paper, Advertising in Asia–From Dependence to Convergence, presented at the AMIC 25th Anniversary Conference at Singapore in June 1996, Katherine Toland Frith, Chair of Advertising, Pennsylvania State University, quoting from Michael Anderson's book, *Madison Avenue in Asia–Politics and Transnational Advertising,* said :

The impact of the transnational corporations (TNCs) on the

needs and aspirations of the Third World masses living amidst scarcity and poverty is not always encouraging. The consumption gap between the few rich and many poor continues. Progress towards reducing the cultural shock, eliminating poverty and other forms of inequality, ensuring citizens full participation in their nation's decision-making process or fostering national self-reliance action has been slow at best.

Caught up in the glamorous world of consumerism (except for miniscule minority of media analysts, academics, conscientious and concerned citizens and NGOs) we are hardly able to assess the impact of this phenomenon. Mandi House has tended to allow the dictates of commercialism to adversely affect the content of its telecasts. The ad blitz which has engulfed the channels of Doordarshan has made many worthwhile programmes, serials and films unwatchable. The actual programme being aired often appears almost fleetingly between waves of commercials. Sometimes ads are deliberately inserted to distort and twist programme sequences. This is against all ethics and norms of public or even commercial telecasting, an unwarranted imposition on viewers. The ads stretch the duration of a film to hours beyond its actual length. In some newspapers, the 'Response Department'. has assumed importance and respectability, which could not have been imagined even a decade earlier. Sometimes one gets the impression that the editorial contents are used only to fill the space left out by commercial and matrimonial advertisements. The result is that in some cases the Response or advertisement department calls the shots and the salary of the advertisment manager compares favourably with, if not more than, that of the editor.

The concept being promoted lately is that selling newspapers is like selling cakes of soap. The paper is a product and the emphasis is on packaging. This means the editor loses authority to the commercial and business side, to modern management methods and techniques, to contract journalism. Does an editor's freedom end where a proprietor's eyebrows begin to rise? Media educator P. N. Malhan asks if the editor is the custodian of the proprietor's vested interests and comments, the value of the newspaper is in its contents, the selection of which is the responsibility of the editor. S. Nihal Singh is pessimistic about the situation in India today. He says : The traditional concept of the editor of a major established English language newspaper was of a man in an ivory tower, disdainful of commerce and the lowly tasks carried out by advertisement and other managers. There was little interaction between such an editor and his circulation manager. The pendulum seems now to have swung to the other extreme in some newspaper establishments. The editor has become a non-entity. The proprietor and his managers lay down the law.

Trivialising Cultural Traditions

Before the disease of privatisation afflicted Doordarshan; programmes were dull but certainly less damaging, at best a waste of a precious national resource. Misconceived privatisation on the other hand has made public broadcasting hostage to crass commercialism. Programmes are not only being drowned in ads but also diverted from the national channels. Doordarshan should develop sensibilities required for public broadcasting which is so far not much in evidence. This is possible only if Doordarshan exploits the advantages of the market instead of becoming its willing victim.

In a seminar organised by AMIC at Kuala Lumpur in August 1995 on, Asian Values in Journalism, the Information Minister of Malaysia in his inaugural address, gave two examples of trivialising our cultural traditions by foreign TV networks and both the examples were from India : The abusive expression used against Mahatma Gandhi in the STAR television programme hosted by Nikki Bedi, and the newsclip of cows straddling the Bombay streets shown on the occasion of the launching of CNN–Doordarshan agreement. In both cases, sensitivities of the Indian viewers were deeply hurt. These instances were not the only one which have evoked concern and unhappiness about the values that the Indian television tends to foster.

Combating Commercialisation of News and Junk Journalism

In order to combat the rising trend of commercialisation of news, McManus has suggested educating journalists to become professionals and to instil in students greater sense of professionalism. Ethical standards are inculcated by journalism educators, by professional groups such as the Radio and the Television News Directors Association and the Society of Professional Journalists, and by mid-career educational experiences such as the Knight and Nieman fellowships. All of these are important actions to improve the quality of journalism, but by themselves, they are likely to prove ineffective.

Interestingly the author admits : A reporter or editor in a profit maximizing media firm, who subordinates market standards to those of journalism, may be tolerated about as much as a counter clerk at McDonald's who refuses to sell fried food. There is simply no evidence that professionalism can restrain corporate demands for profit. Bagdikian recently pointed out that American news media have never had better educated or more professional journalists than they have today, yet bottom-line oriented executives are increasingly moving their organizations away from quality journalism.

Whatever may be the compulsions of catering to the interests of investors and advertisers, efforts to curb junk journalism through an emphasis on professionalisation, both in journalism training and in organization of reporters and editors should continue. Increasing public awareness of the consequences of junk journalism may be the surest route to a healthier information diet. There lies some hope in combating the rising influence of market forces on media institutions.

The concern about the role of market forces in encouraging *paparazzi*–a variant of junk journalism–was deep and widespread in the wake of the death of princess Diana in a car crash in Paris last year. So much has been already written about *paparazzi,* which began as an Italian phenomenon, "pesty shutterbugs flitting through the streets of Rome", endangering life and limb--in search of celebrity snaps—that it need not be expanded here; however, the role of the tabloid press and the relentless profit motive which derives it towards junk journalism–or gutter press if you like–needs to be noted.

Named after a character in the 60s classic movie, "La Dolca Vita" *Paparazzi* are mostly freelance photographers who believed that : *earnings from a single photograph, could turn them from paupers to men who never needed to work again.* Strongly motivated by quick gains from easy money. On the other hand, the *paparazzi,* however, sought refuge in the justification that millions of people bought the newspapers that carried their stuff. "We only give the public what it wants", is the commonest refrain, the one we are used to, when some one draws attention to the vulgarity and low taste being traded off in the media in return for higher circulations and enhanced office earnings.

In the forefront of the process of creating a highly lucrative global market was *The Sun,* with a daily circulation of 3.8 million, and a part of Rupert Murdoch's news and entertainment empire in England.

Since the arrival of Rupert Murdoch on Fleet Street in 1970s, competition at the tabloid end of the newspaper market, has completely metamorphosied the tone and content in the British press. With Rupert Murdoch's entry, the British media–beside *The Sun,* the Murdoch empire comprises *The Times* in London, *The New York Post,* the *Australian,* Fox TV in US and STAR TV in Asia, as also the publishing conglomerate *Harper Collins*–tends to be treated as a part of insatiable global market place.

The Australian born media tycoon has been often accused of brazen political partisanship in the affairs of countries where his media empire operates. He switched his *Sun* newspaper to support labour before the general elections in England last year because of a reported understanding between him and the British Prime Minister Tony Blair. More recently the world press had the opportunity to examine closely and report the clumsy way *Harper Collins* cancelled the publication of Chris Pattens memoirs, for Murdoch to ingratiate himself with the Chinese–in the hope of getting access to the vast market opportunities which China offers. Murdoch had earlier dropped BBC World Service television from the STAR TV service by satellite to China and India, since BBC's reporting Asia did not meet with approval by the Chinese Government. In the recent General Elections in India, the blatantly partisan role of the STAR News Channel, in projecting Congress as the winning party has been widely noticed and commented. Murdoch is increasingly using the clout of the Indian arm of the media empire– STAR News Channel–as a political player would do. Any Broadcast Policy would do well to take note of the potential and propensity for interference in India's internal affairs by the foreign satellite channels. All these trends–growth of *paparazzi* as a result of pursuing profit motives, blatant interference in the politics of the host country, cancelling the publishing contract–should derive home the lesson for us in India that it would be highly detrimental for any government to reverse the long standing national policy of ban on the entry of foreign press. Equally important would be to restrict the equity for foreign broadcasting organizations to 20 per cent.

Corporatisation of Media

The rapid technological advances have brought about far-reaching changes, leading to wider opportunity of investment for transnational corporate capital. The transnational and multinational corporations are actively utilising media technology for profit motives–one example being that of Rupert Murdoch's STAR TV business in the

Asian region through the application of satellite technology in commercial broadcasting.

Media–including the Press, which had such a glorious role to play in the Indian Freedom Movement, and the Radio and Television (the social service mandate) are no longer a service; they are becoming more and more of industry and commercial ventures. In this context, it will be worthwhile to reproduce one paragraph from the essay–Open Markets and the Future of Television, contributed by the editors of the book–*Contemporary Television–Eastern Perspectives,* David French and Michael Richards :

> Satellite television, with its ownership in the hands of buccaneering capitalists, has then been a crucial factor in this process of change. It is crucial to understand that the process of technological convergence has not just been a matter of changing technology but also of the importation into the domain of television of vigorously commercial approaches. If satellites are associated with names such as Ted Turner and Rupert Murdoch, then the business of computing and digital technology resonates as strongly with those of Bill Gates and his younger contemporaries. Crucially, the industries and technologies with which broadcasting is converging have always been led by private capital and have only in aberrant national circumstances been subject to regulation for the public good. In other words, if countries wish to jump on the broadband highway, collectively to ride the Internet and the Worldwide Web, then they have superficially at least, to speak its language and that is the language of private capital and of the free market.

Regional Language Press

Apart from the growth of regional language press, two significant emerging trends in the Indian context need to be mentioned—the ongoing movement for the Right to Information and the increasing involvement of the media NGOs in impacting policy formulation.

The phenomenal expansion of the language press has brought about a marked change the way public issues are discussed and conducted. Over the past decade or so the language press has gained enviable power over local politics so much so that regional power centres are often seen wooing it in an effort to gain its support at the local level. The emphasis has also shifted from the national to regional politics. The editors and reporters of language press are no longer considered *shudras* in the media hierarchy. According to one reckoning, the total circulation of Hindi dailies in 1952 was 3.8 lakhs (76 dailies), in 1991 the number of dailies was 1182. The data supplied by 426 dailies gave the circulation figures as 81 lakhs. By 1990, the number of Indian language dailies with more than one lakh circulation rose to 36 and more than half a dozen of these commanded circulation of over three lakhs. All this has added new dimensions to the contemporary media landscape in the country making it more pluralistic, diverse and decentralised.

The Right to Know Campaign

In one of the significant milestone events as a part of the Right to Know campaign, the National Institute of Rural Development in Hyderabad hosted a National Workshop in September 1997. While the Workshop recommended that apart from passing of the 'Right to Know' bill, several outdated acts and manuals including the Official Secrets Acts, Conduct Rules for Civil Servants, Public Records Act be amended, it also observed : Fifty years after the British left this country, governments are hostile to any attempts by members of the public to pry into their affairs. While part of the difficulty lies in the attitudes of the bureaucracy, the political system has not been in favour of openness and transparency either. Secrecy is the watchword in governmental affairs–and the 74 years old Official Secrets Act, 1923 is the sacred text.

Earlier than the Hyderabad Workshop, the Right to Know campaign had metamorphosied into a movement from grassroots embedded in the protest by the villagers in several districts in Rajasthan. The movement was to ensure accountability in development works. The struggle for Right to Information was essentially to obtain photocopies of bills, receipts, muster rolls and other documents relating to the above development related project accounts. The Mazdoor, Kisan Shakti Sangathan (MKSS) has been in the vanguard of the struggle in Beawar area of Rajasthan since 1990. In due courses, several leading media personalities and organisations including Press Council of India, Press Institute of India, National Media Centre joined hands to create a strong public opinion for enacting the bill for Right to Know. However, the bureaucracy in India is nothing if not retrograde and stumbling block in fulfilling this most notable objective. When MKSS organised several *jan sunwayis* (open public hearings), where hundreds of the very poorest from far off villages came to speak, listen and share their experiences, alarm bells started ringing, official secrets the close preserve of the bureaucracy came under threat and efforts to stonewall the initiative of getting access to information put in place.

At present there appears to be stalemate in getting any legislation for access to information through. May be the movement needs relentless efforts and strong public opinion to push the legislation forward.

Media NGOs and Policy Formulation

Another notable development in the field of media education, research and policy formulation has been the significant contribution made by several media NGOs. The foremost of these organisations is the Singapore-based Asian Media Information and Communication Centre– AMIC. Established in 1971, the Centre functions as a clearing house of information, focal point for communication practitioners and scholars to exchange experiences and ideas, and as a publishing house devoted to communication issues.

The story of AMIC is the story of an organisation reaching out not only with information about mass communication in the region but also with teaching, training and research inputs, aiding in a greater appreciation and use of mass communication as a tool to help the countries of Asia face the challenges and avail of the opportunities of the information age.

In this contest Vijay Menon, the AMIC Secretary General observes: The story of AMIC began in 1971. It was set up primarily as a regional documentation centre in response to a suggestion from UNESCO. A graceful building on Newton Road was made available to the Centre and AMIC was in business. Mass communication documentation is still an important activity at AMIC. But the range of activities has grown. AMIC has been fortunate in getting both moral and material support from the Government of Singapore and the Friedrich Ebert Stiftung, an independent foundation in the Federal Republic in Germany. UNESCO which has conferred on AMIC recognition as a Category 'B' NGO, continues to provide valuable encouragement and assistance. Specific projects are funded by various organisations around the world. It is non-governmental organisation and a non-profit foundation.

AMIC has initiated and coordinated various national and regional applied communications research projects. Of AMIC's role in research, the distinguished scholar, John Lent, had this to say :

> Mass communication made a slow start in Southeast Asia, as it generally had been in the Third World. . . . Virtually all aspects of communication had gone unstudied. The boon to the development of mass communication as a field of study, not only in Southeast Asia but all of Asia was the Asian Mass Communication Research and Information Centre founded in Singapore in 1971 . . . The role of research has been featured in nearly all AMIC seminars and conferences.

AMIC's publications programme produces *Media Asia,* a quarterly for media practitioners, *Asian Journal of Communication,* a biannual scholarly journal, and *Asian Mass Communication Bulletin,* a newsletter containing news and information on current media trends as well as bulletins on AMIC activities. In addition, the publications unit publishes books, monographs and occasional papers.

For the past several years, AMIC has introduced some very successful programmes; both at the Asia Pacific level–Annual Conference which draws galaxy of communication professionals, administrators and policy makers; as also incountry programmes where participants drawn from diverse disciplines : print journalism, mass communication, extension education, electronic media, sociology interact in seminars and conferences.

In India, the Bangalore based non-profit trust *Voices* has been working on several aspects of communication related activities including training in communication skills, consultations and workshops on emerging perspectives and issues, production of communication material, studies related to media and cultural forms and advocacy and networking with action groups. Its Women's Health Resource Centre disseminates media material on women's health. The journal *Voices* provides a forum to a vast cross section of media academics and scholars, workers in the voluntary and social sections and all others interested in working for change in the society. Headed by Sucharita Eashwar, *Voices'* initiative in conceptualising Community Radio experiment in India is a step in the direction of encouraging participatory communication.

The Delhi-based media organization, National Media Centre, has been promoting interaction both between journalists at home and with scribes abroad. Over the years it has been providing opportunities for continuing interaction on media related subjects–Right to Know, Media Bureaucracy and Judiciary; Pressures on Media, Media and Peace, Education for All and Entry of Foreign

Press in India. The General Secretary, Upendra Vajpeyi, who initiates and coordinates these programmes, has several prospective publications lined up : Birth and Growth of Journalism in Indian languages, Role of Press in Freedom Movement, and Women in India.

The brief note about the working of these media NGOs, by no means completes reference to such organisations. There would certainly be many more such institutions whose contribution need to be further explored. As we prepare to enter the twentyfirst century we would be needing many more such organisations, particularly so when the shifting paradigms, and emerging concepts add new perspectives to the media landscape in India and abroad. These developments and trends need to be continually monitored, analysed and documented.

An other effort, in due course of time, perhaps is called for.

ANNEXURE-I

THE PRASAR BHARATI (BROADCASTING CORPORATION OF INDIA) ACT, 1990
NO. 25 OF 1990

(12th September, 1990)

An Act to provide for the establishment of a Broadcasting Corporation for India, to be known as Prasar Bharati, to define its composition, functions and powers and to provide for matters connected therewith or incidental thereto.

Be it enacted by Parliament in the Forty-first Year of the Republic of India as follows :

CHAPTER I

PRELIMINARY

Short title, extent and Commencement.

1. (1) This Act may be called the Praṣar Bharati (Broadcasting Corporation of India) Act, 1990.

(2) It extends to the whole of India.

(3) It shall come into force on such date as the Central Government may, by notification, appoint.

Definitions.

2. In this Act, unless the context otherwise requires—

(a) "Akashvani" means the offices, stations and other

The Prasar Bharati (Broadcasting Corporation of India) Act, was enacted by the Parliament in September 1990. Seven years after it was enacted, it was notified only on 22 July 1997. It came into force from 15 September 1997. An Ordinance amending the Act was promulgated by the President of India on 30 October 1997, making far reaching changes in the Act. With the changeover in the Government, it is envisaged that the Ordinance will be allowed to lapse. The Government also proposes to retain the Parliamentary Committee, as envisaged in Section 13(1) to oversee the working of the Corporation.

establishments, by whatever name called, which, immediately before the appointed day, formed part of or were under the Director-General, All-India Radio of the Union Ministry of Information and Broadcasting;

(b) "appointed day" means the date appointed under Section 3;

(c) "broadcasting" means the dissemination of any form of communication like signs, signals, writing, pictures, images and sound of all kinds by transmission of electro-magnetic waves through space or through cables intended to be received by the general public either directly or indirectly through the medium of relay stations and all its grammatical variations and cognate expressions shall be construed accordingly.

(d) "Board" means the Prasar Bharati Board;

(e) "Broadcasting Council" means the Council established under Section 14;

(f) "Chairman" means the Chairman of the Corporation appointed under Section 4;

(g) "Corporation" means the Prasar Bharati (Broadcasting Corporation of India) established under Section 3;

(h) "Doordarshan" means the offices, kendras and other establishments, by whatever name called, which, immediately before the appointed day, formed part, of or were under the Directorate-General, Doordarshan of the Union Ministry of Information and Broadcasting;

(i) "elected Member" means a Member elected under section 3;

(j) "Executive Member" means the Executive Member appointed under Section 4;

(k) "kendra" means any telecasting centre with studios or transmitters or both and includes a relay station;

(l) "Member" means a Member of the Board;

(m) "Member (Finance)" means the Member (Finance) appointed under Section 4;

(n) "Member (Personnel)" means the Member (Personnel) appointed under Section 4;

(o) "Nominated Member" means the Member nominated by the Union Ministry of Information and Broadcasting under Section 3;

(p) "Non-lapsable Fund" means the Fund created from the commercial revenues of Akashvani and Doordarshan to meet expenditure on certain schemes;

(q) "notification" means a notification published in the Official Gazette;

(r) "Part-time Member" means a Part-time Member of the Board appointed under Section 4, but does not include an ex-officio Member, the Nominated Member or an elected Member;

(s) "prescribed" means prescribed by rules made under this Act;

(t) "Recruitment Board" means a board established under Sub-section (i) of Section 10;

(u) "regulations" means regulations made by the Corporation under this Act;

(v) "station" means any broadcasting station with studios or transmitters or both and includes a relay station;

(w) "Whole-time Member" means the Executive Member, Member (Finance) or Member (Personnel);

(x) "year" means the financial year.

CHAPTER II

PRASAR BHARATI (BROADCASTING CORPORATION OF INDIA)

Establishment and Composition of Corporation

3. (1) With effect from such date as the central government may by notification appoint in this behalf, there shall be established for the purposes of this Act a Corporation, to be known as the Prasar Bharati (Broadcasting Corporation of India).

(2) The Corporation shall be a body corporate by the name aforesaid, having perpetual succession and a common seal with power to acquire, hold and dispose of property, both movable and immovable, and to contract, and shall by the said name sue and be sued.

(3) The headquarters of the Corporation shall be at New Delhi and the Corporation may establish offices, kendras or stations at other places in India and, with the previous approval of the Central Government, outside India.

(4) The general superintendence, direction and management of the affairs of the Corporation shall vest in the Prasar Bharati Board which may exercise all such powers and do all such acts and things as may be exercised or done by the Corporation under this Act.

(5) The Board shall consist of–

(a) a Chairman;

(b) one Executive Member;

(c) one Member (Finance);

(d) one Member (Personnel);
(e) six Part-time Members;
(f) Director-General (Akashvani), ex-officio;
(g) Director General (Doordarshan); ex-officio;
(h) one representative of the Union Ministry of Information and Broadcasting, to be nominated by that Ministry; and
(i) two representatives of the employees of the Corporation, of whom one shall be elected by the engineering staff from amongst themselves and one shall be elected by the other employees from amongst themselves.

(6) The Corporation may appoint such committees as may be necessary for the efficient performance, exercise and discharge of its functions, power and duties.
Provided that all or a majority of the members of each committee shall be Members and a member of any such committee who is not a Member shall have only the right to attend meetings of the committee and take part in the proceedings thereof, but shall not have the right to vote.

(7) The Corporation may associate with itself, in such manner and for such purposes as may be provided by regulations, any person whose assistance or advice it may need in complying with any of the provisions of this Act and a person so associated shall have the right to take part in the discussions of the Board relevant to the purposes for which he has been associated, but shall not have the right to vote.

(8) No act or proceeding of the Board or any committee appointed by it under sub-section (6) shall be invalidated merely by reason of–
(a) any vacancy in, or any defect in the constitution of, the Board or such committee, or
(b) any defect, in the appointment of a person acting as a Member or a member of such committee; or
(c) any irregularity in the procedure of the Board or such committee not affecting the merits of the case.

Appointment of Chairman and other Members

4. (1) the Chairman and the other Members, except the ex-officio members, the Nominated Member and the elected Members shall be appointed by the President of India on the recommendation of a committee consisting of
(a) the Chairman of the Council of States, who shall be the

Chairman of the Committee;

(b) the Chairman of the Press Council of India established under Section 4 of the Press Council Act, 1978; and

(c) one nominee of the President of India.

(2) No appointment of a Member shall be invalidated merely by reason of any vacancy in, or any defect in the constitution of, the committee appointed under sub-section (1).

(3) The Chairman and the Part-time Members shall be persons of eminence in public life; the Executive Member shall be a person having special knowledge or practical experience in respect of such matters as administration, management, broadcasting, education, literature, culture, arts, music, dramatics or journalism, the Member (Finance) shall be a person having special knowledge or practical experience in respect of financial matters and the Member (Personnel) shall be a person having special knowledge or practical experience in respect of personnel management and administration.

(4) The recommendations made by the committee constituted under sub-section (1) shall be binding for the purposes of appointment under this section.

Powers and Functions of Executive Member

5. The Executive Member shall be the Chief Executive of the Corporation and shall, subject to the control and supervision of the Board, exercise such powers and discharge such functions of the Board as it may delegate to him.

Term of Office, Conditions of Service, etc. of Chairman and other Members

6. (1) The Chairman shall be Part-time Member and shall hold office for a term of six years from the date on which he enters upon his office.

(2) The Executive Member, the Member (Finance) and the Member (Personnel) shall be whole-time Members and every such Member shall hold office for a term of six years from the date on which he enters upon his office or until he attains the age of sixty-two years, whichever is earlier.

(3) The term of office of Part-time Members shall be six years, but one-third of such Members shall retire on the expiration of every second year.

(4) The term of office of an elected Member shall be two years or till he ceases to be an employee of the Corporation, whichever is earlier.

(5) As soon as may be after the establishment of the Corporation, the President of India may, by order, make such provision as he thinks fit for curtailing the term of office of some of the Part-time Members then appointed in order that one-third of the Members holding office as such Part-time Members shall retire in every second year thereafter.

(6) Where before the expiry of the term of office of a person holding the office of Chairman, or any other Member, a vacancy arises, for any reason whatsoever, such vacancy shall be deemed to be a casual vacancy and the person appointed or elected to fill such vacancy shall hold office for the unexpired period of the term for which his predecessor in office would have held office if such vacancy had not arisen.

(7) The Whole-time Members shall be the employees of the Corporation and as such shall be entitled to such salaries and allowances and shall be subject to such conditions of service in respect of leave, pension (if any), provident fund and other matters as may be prescribed :

Provided that the salaries and allowances and the conditions of service shall not be varied to their disadvantage after their appointment.

(8) The Chairman and Part-time Members shall be entitled to such allowances as may be prescribed.

Removal and Suspension of Chairman and Members

7. (1) Subject to the provisions of sub-section (3), the Chairman or any other Member, except an ex-officio Member, the Nominated Member and an elected Member, shall only be removed from his office by order of the President of India on the ground of misbehaviour after the Supreme Court, on a reference being made to it by the President, has, on inquiry held in accordance with such procedure as the Supreme Court may by rules provide, reported that the Chairman or such other Member, as the case may be ought on such ground, be removed.

(2) The President may suspend from office the Chairman or other Member, except an ex-officio Member, the Nominated Member or an elected Member, in respect of whom a reference has been made to the Supreme Court under sub-section (1) until

the President has passed orders on receipt of the report of the Supreme Court on such reference.

(3) Notwithstanding anything contained in sub-section (1), the President may, by order, remove the Chairman or any Whole-time Member from his office if such Chairman or such whole-time Member–

(a) ceases to be a citizen of India; or

(b) is adjudged an insolvent; or

(c) engages during his term of office in any paid employment outside the duties of his office; or

(d) is convicted of any offence involving moral turpitude; or

(e) is, in the opinion of the President, unfit to continue in office by reason of infirmity of body or mind;

Provided that the President may, by order, remove any Part-time Member from his office if he is adjudged an insolvent or is convicted of any offence involving moral turpitude or where he is, in the opinion of the President, unfit to continue in office by reason of infirmity of body or mind.

(4) If the Chairman or any Whole-time Member, except any ex-officio Member, the Nominated Member or any elected Member, is, or becomes in any way concerned or interested in any contract or agreement made by or on behalf of the Corporation or the Government of India or the Government of a State or, participates in any way in the profit thereof, or in any benefit or emolument arising therefrom than as a member, and in common with other members of an incorporated company, he shall, for the purposes of sub-section (1), be deemed to be guilty of misbehaviour.

(5) If a Part-time Member is, or becomes in any way concerned, or interested in any contract or agreement made by or on behalf of the Corporation, he shall, for the purposes of sub-section (1), be deemed to be guilty of misbehaviour.

(6) The Chairman or any other Member may resign his office by giving notice thereof in writing to the President of India and on such resignation being accepted, the Chairman or other Member shall be deemed to have vacated his office.

Meetings of Board

8. (1) The Board shall meet at such times and places and shall observe

such rules of procedure in regard to the transaction of business at its meetings (including the quorum of meetings) as may be provided by regulations :

Provided that there shall not be less than six meetings every year but three months shall not intervene between one meeting and the next meeting.

(2) A Member shall be deemed to have vacated his office if he absents himself for three consecutive meetings of the Board without the leave of the Chairman.

(3) The Chairman shall preside at the meetings of the Board and if for any reason he is unable to attend any meeting, the Executive Member and in the absence of both any other Member elected by the Members present at such meeting shall preside at the meeting.

(4) All questions which come up before any meeting of the Board shall be decided by a majority of the votes of the Members present and voting and, in the event of an equality of votes, the Chairman, or in his absence, the person presiding, shall have and exercise a second or casting vote.

Officers and other Employees of Corporation

9. (1) Subject to such control, restrictions and conditions as may be prescribed, the Corporation may appoint, after consultation with the Recruitment Board, the Director-General (Akashvani), the Director General (Doordarshan) and such other officers and other employees as may be necessary.

(2) The method of recruitment of such officers and employees and all other matters connected therewith and the conditions of service of such officers and other employees shall be such as may be provided by regulations.

Establishment of Recruitment Boards

10. (1) The Corporation shall as soon as may be, after the appointed day and in such manner and subject to such conditions and restrictions as may be prescribed, establish for the purposes of section 9, one or more Recruitment Boards consisting wholly of persons other than the Members, officers and other employees of the Corporation;

Provided that for the purposes of appointment to the posts carrying scales of pay which are not less than that of a Joint

Secretary to the Central Government, the Recruitment Board shall consist of the Chairman, other Members, the ex-officio Members, the Nominated Member and the elected Members.

(2) The qualifications and other conditions of service of the members constituting the Recruitment Board and the period for which such member shall hold office, shall be such as may be prescribed.

Transfer of Service of Existing Employees to Corporation

11\. (1) Where the Central Government has ceased to perform any functions which under section 12 are the functions of the Corporation, it shall be lawful for the Central Government to transfer, by order and with effect from such date or dates as may be specified in the order, to the Corporation any of the officers or other employees serving in the Akashvani or Doordarshan and engaged in the performance of those functions:

Provided that no order under this sub-section shall be made in relation to any officer or other employee in the Akashvani or Doordarshan who has, in respect of the proposal of the Central Government to transfer such officer or other employee to the Corporation, intimated with such time as may be specified in this behalf by the Central Government, his intention of not becoming an employee of the Corporation.

(2) The provisions of sub-section (1) shall also apply to the members of the Indian Information Service, the Central Secretariat Service or any other service or to persons borne on cadres outside Akashvani and Doordarshan who have been working in Akashvani or Doordarshan immediately before the appointed day :

Provided that where any such member intimates, within the time specified in sub-section (1) his intention of not becoming an employee of the Corporation but to continue on deputation, he may be allowed to continue on deputation in accordance with such terms and conditions as may be prescribed.

(3) In making an order under sub-section (1), the Central Government shall, as far as may be, take into consideration the functions which the Akashvani or, as the case may be, Doordarshan has ceased or ceases to perform and the area in which such functions have been or are performed.

(4) An officer or other employee transferred by an order under sub-section (1) shall, on and from the date of transfer, cease to

be an employee of the Central Government and become an employee of the Corporation with such designation as the Corporation may determine and shall, subject to the provisions of sub-sections (5) and (6), be governed by such regulations as may be made as respects remuneration and other conditions of service including pension, leave and provident fund and shall continue to be an officer or other employee of the Corporation unless and until his employment is terminated by the Corporation.

(5) Every officer or other employee transferred by an order made under sub-section (1) shall, within six months from the date of transfer, exercise his option, in writing, to be governed—

(a) by the scale of pay applicable to the post held by him in the Akashvani or Doordarshan immediately before the date of transfer or by the scale applicable to the post under the Corporation to which he is transferred;

(b) by the leave, provident fund, retirement or other terminal benefits admissible to employees of the Central Government in accordance with the rules or orders of the Central Government, as amended from time to time, or the leave, provident fund or other terminal benefits admissible to the employees of the Corporation under the regulations and such option once exercised under this Act shall be final :

Provided that the option exercised under clause (a) by an officer or other employee shall be applicable only in respect of the post under the Corporation to which such officer or other employee is transferred and on appointment to a higher post under the Corporation he shall be eligible only for the scale of pay applicable to such higher post :

Provided further that if immediately before the date of his transfer any such officer or other employee is officiating in a higher post under the Government either in a leave vacancy or any other vacancy of a specified duration, his pay on transfer shall be protected for the unexpired period of such vacancy and thereafter he shall be entitled to the scale of pay applicable to the post under the Government to which he would have reverted or to the scale of pay applicable to the post under the Corporation to which he is transferred, whichever he may opt.

Provided also that when an officer or other employee serving in the Union Ministry of Information and Broadcasting or in any of its attached or subordinate offices is promoted to officiate in a higher post in the Ministry or office subsequent to the transfer to the Corporation of any other officer or employee senior to him in that Ministry or office before such transfer, the officer or other employee who is promoted to officiate in such higher post shall, on transfer to the Corporation, be entitled only to the scale of pay applicable to the post he would have held but for such promotion or the scale of pay applicable to the post under the Corporation to which he is transferred whichever he may opt.

(6) No officer or other employee transferred by an order made under sub-section (1) or sub-section (2) –

(a) shall be dismissed or removed by an authority subordinate to that competent to make a similar or equivalent appointment under the Corporation as may be specified in the regulations; and

(b) shall be dismissed or removed or reduced in rank except after an inquiry in which he has been informed of the charges against him and given a reasonable opportunity of being heard in respect of those charges :

Provided that where it is proposed after such inquiry to impose upon him any such penalty, such penalty may be imposed on the basis of evidence adduced during such inquiry and it shall not be necessary to give such person an opportunity of making representation on the proposed penalty.

Provided further that clause (b) shall not apply where an officer or other employee is dismissed or removed or reduced in rank on the ground of conduct which has led to his conviction on a criminal charge.

Functions and Powers of Corporation

12. (1) Subject to the provisions of this Act, it shall be the primary duty of the Corporation to organise and conduct public broadcasting services to inform, educate and entertain the public and to ensure a balanced development of broadcasting on radio and television.

Explanation–For the removal of doubts, it is hereby declared that the provisions of this section shall be in addition to, and not in derogation of the provisions of the Indian Telegraph Act, 1885.

(2) The Corporation shall, in the discharge of its functions be

guided by the following objectives, namely :

(a) upholding the unity and integrity of the country and the values enshrined in the Constitution;

(b) safeguarding the citizen's right to be informed freely, truthfully and objectively on all matters of public interest, national or international, and presenting a fair and balanced flow of information including, contrasting views without advocating any opinion or ideology of its own;

(c) paying special attention to the fields of education and spread of literacy, agriculture, rural development, environment, health and family welfare and science and technology;

(d) providing adequate coverage to the diverse cultures and languages of the various regions of the country by broadcasting appropriate programmes;

(e) providing adequate coverage to sports and games so as to encourage healthy competition and the spirit of sportsmanship;

(f) providing appropriate programmes keeping in view the special needs of the youth;

(g) informing and stimulating the national consciousness in regard to the status and problems of women and paying special attention to the upliftment of women;

(h) promoting social justice and combating exploitation, inequality and such evils as untouchability and advancing the welfare of the weaker sections of the society;

(i) safeguarding the rights of the working classes and advancing their welfare;

(j) serving the rural and weaker sections of the people and those residing in border regions, backward or remote areas;

(k) providing suitable programmes keeping in view the special needs of the minorities and tribal communities;

(l) taking special steps to protect the interests of children, the blind, the aged, the handicapped and other vulnerable sections of the people;

(m) promoting national integration by broadcasting in a manner that facilitates communication in the languages in India; and facilitating the distribution of regional broadcasting services in every state in the languages of that state;

(n) providing comprehensive broadcast coverage through the choice of appropriate technology and the best utilisation of the broadcast frequencies available and ensuring high quality reception;

(o) promoting research and development activities in order to ensure that radio and television broadcast technology are constantly updated; and

(p) expanding broadcasting facilities by establishing additional channels of transmission at various levels.

(3) In particular, and without prejudice to the generality of the foregoing provisions, the Corporation may take such steps as it thinks fit –

(a) to ensure that broadcasting is conducted as a public service to provide and produce programmes;

(b) to establish a system for the gathering of news for radio and television;

(c) to negotiate for purchase of, or otherwise acquire, programmes and rights or privileges in respect of sports and other events, films, serials, occasions, meetings, functions or incidents of public interest, for broadcasting and to establish procedures for the allocation of such programmes, rights or privileges to the services;

(d) to establish and maintain a library or libraries of radio, television and other materials;

(e) to conduct or commission, from time to time, programmes, audience research, market or technical service, which may be released to such persons and in such manner and subject to such terms and conditions as the Corporation may think fit;

(f) to provide such other services as may be specified by regulations.

(4) Nothing in sub-section (2) and (3) shall prevent the Corporation from managing on behalf of the Central Government and in accordance with such terms and conditions as may be specified by that Government the broadcasting of External Services and monitoring of broadcasts made by organisations outside India on the basis of arrangements made for reimbursement of expenses by the Central Government.

(5) For the purposes of ensuring that adequate time is made available for the promotion of the objectives set out in this section, the Central Government shall have the power to determine the maximum limit of broadcast time in respect of the advertisement.

(6) The Corporation shall be subject to no civil liability on the ground merely that it failed to comply with any of the provisions of this section.

(7) The Corporation shall have power to determine and levy fees and other service charges for or in respect of the advertisements and such programmes as may be specified by regulations :

Provided that the fees and other service charges levied and collected under this sub-section shall not exceed such limits as may be determined by the Central Government, from time to time.

Parliamentary Committee

13. (1) There shall be constituted a Committee consisting of twenty-two members of Parliament, of whom fifteen from the House of the People to be elected by the Members thereof and seven from the Council of States to be elected by the Members thereof in accordance with the system of proportional representation by means of the single transferable vote, to oversee that the Corporation discharges its functions in accordance with the provisions of this Act and, in particular, the objectives set out in Section 12 and submit a report thereon to Parliament.

(2) The Committee shall function in accordance with such rules as may be made by the Speaker of the House of the people.

Establishment of Broadcasting Council, Term of Office and Removal, etc. of Members Thereof

14. (1) There shall be established, by notification, as soon as may be after the appointed day, a Council, to be known as the Broadcasting Council, to receive and consider complaints referred to in Section 15 and to advise the Corporation in the discharge of its functions in accordance with the objectives set out in Section 12.

(2) The Broadcasting Council shall consist of–

(i) a President and ten other members to be appointed by the President of India from amongst persons of eminence in public life;

(ii) four Members of Parliament, of whom two from the House of the People to be nominated by the Speaker thereof and two from the Council of States to be nominated by the Chairman thereof.

(3) The President of the Broadcasting Council shall be a whole-time member and every other member shall be a part-time member and the President or the part-time member shall hold

office as such for a term of three years from the date on which he enters upon his office.

(4) The Broadcasting Council may constitute such number of Regional Councils as it may deem necessary to aid and assist the Council in the discharge of its functions.

(5) The President of the Broadcasting Council shall be entitled to such salary and allowances and shall be subject to such conditions of service in respect of leave, pension (if any), provident fund and other matters as may be prescribed:

Provided that the salary and allowances and the conditions of service shall not be varied to the disadvantage of the President of the Broadcasting Council after his appointment.

(6) The other members of the Broadcasting Council and the members of the Regional Councils constituted under sub-section (4) shall be entitled to such allowances as may be prescribed.

Jurisdiction of and the Procedure to be followed by, Broadcasting Council

15. (1) The Broadcasting Council shall receive and consider complaints from–

(i) any person or group of persons alleging that a certain programme or broadcast or the functioning of the Corporation in specific cases or in general is not in accordance with the objectives for which the Corporation is established;

(ii) any person (other than an officer or employee of the Corporation) claiming himself to have been treated unjustly or unfairly in any manner (including unwarranted invasion of privacy, misrepresentation, distortion or lack of objectively) in connection with any programme broadcast by the Corporation.

(2) A complaint under sub-section (1) shall be made in such manner and within such period as may be specified by regulations.

(3) The Broadcasting Council shall follow such procedure as it thinks fir for the disposal of complaints received by it.

(4) If the complaint is found to be justified either wholly or in part, the Broadcasting Council shall advise the Executive Member to take appropriate action.

(5) If the Executive Member is unable to accept the recommenda-

tion of the Broadcasting Council, he shall place such recommendation before the Board for its decision thereon.

(6) If the Board is also unable to accept the recommendation of the Broadcasting Council, it shall record its reasons therefore and inform the Broadcasting Council accordingly.

(7) Notwithstanding anything contained in sub-section (5) and (6), where the Broadcasting Council deems it appropriate, it may, for reasons to be recorded in writing, require the Corporation to broadcast its recommendations with respect to a complaint in such manner as the Council may deem fit.

CHAPTER III

ASSETS, FINANCES AND ACCOUNTS

Transfer of Certain Assets, Liabilities, etc. of Central Government to Corporation

16. As from the appointed day–

(a) all property and assets (including the Non-lapsable Fund) which immediately before that day vested in the Central Government for the purpose of Akashvani or Doordarshan or both shall stand transferred to the Corporation on such terms and conditions as may be determined by the Central Government and the book value of all such property and assets shall be treated as the capital provided by the Central Government to the Corporation.

(b) all debts, obligations and liabilities incurred, all contracts entered into and all matters and things engaged to be done by, with or for the Central Government immediately before such day for or in connection with the purposes of Akashvani or Doordarshan or both shall be deemed to have been incurred, entered into and engaged to be done by, with or for the Corporation;

(c) all sums of money due to the Central Government in relation to the Akashvani or Doordarshan or both immediately before such day shall be deemed to be due to the Corporation;

(d) all suits and other legal proceedings instituted or which could have been instituted by or against the Central Government immediately before such day for any matter in relation to the Akashvani or Doordarshan or both may be continued or instituted by or against the Corporation.

Grants, etc. by Central Government

17. For the purposes of enabling the Corporation to discharge its function efficiently under this Act, the Central Government may, after due appropriation made by Parliament by law in this behalf, pay to the Corporation in each financial year–
    (i) the proceeds of the broadcast receiver licence fees, if any, as reduced by the collection charges; and
    (ii) such other sums of money as that Government considers necessary by way of equity, grant-in-aid or loan.

Fund of Corporation

18. (1) The Corporation shall have its own fund and all the receipts of the Corporation (including the amounts which stand transferred to the Corporation under Section 16) shall be credited to the Fund and all payments by the Corporation shall be made therefrom.

(2) All moneys belonging to the Fund shall be deposited in one or more nationalised banks in such manner as the Corporation may decide.

(3) The Corporation may spend such sums as it thinks fit for performing its functions under this Act and such sums shall be treated as expenditure payable out of the Fund of the Corporation.

Explanation—For the purposes of this section, "nationalised bank" means a corresponding new bank specified in the First Schedule to the Banking Companies (Acquisition and Transfer of Undertakings) Act, 1970 or a corresponding new bank specified in the First Schedule to the Banking Companies (Acquisition and Transfer of Undertakings) Act, 1980.

Investment of Moneys

19. The Corporation may invest its moneys in the securities of the Central Government or any State Government or in such other manner as may be prescribed.

Annual Financial Statement of the Corporation

20. (1) The Corporation shall prepare, in each financial year, an Annual Financial Statement for the next financial year showing

separately–

(a) the expenditure which is proposed to be met from the internal resources of the Corporation; and

(b) the sums required from the Central Government to meet other expenses, and distinguishing—

(i) revenue expenditure from other expenditure; and

(ii) non-plan expenditure from plan expenditure.

(2) The Annual Financial Statement shall be prepared in such form and forwarded at such time to the Central Government for its approval as may be agreed to by that Government and the Corporation.

Accounts and Audit of Corporation

21. (1) The Corporation shall maintain proper accounts and other relevant records and prepare an annual statement of accounts in such form and in such manner as may be prescribed.

(2) The accounts of the Corporation shall be audited by the Comptroller and Auditor-General of India at such intervals as may be specified by him and any expenditure incurred in connection with such audit shall be payable by the Corporation to the Comptroller and Auditor-General.

(3) The Comptroller and Auditor-General and any person appointed by him in connection with the audit of the accounts of the corporation shall have the same rights and privileges and authority in connection with such audit as the Comptroller and Auditor-General has in connection with the audit of Government accounts and, in particular, shall have the right to demand the production of books, accounts, connected vouchers and other documents and papers and to inspect any of the offices of the Corporation.

(4) The accounts of the Corporation as certified by the Comptroller and Auditor-General of India or any other person appointed by him in this behalf together with the audit report thereon shall be forwarded annually to the Central Government and that Government shall cause the same to be laid before each House of Parliament.

Corporation not Liable to be Taxed

22. Notwithstanding anything contained in the Income-tax Act, 1961, or any other enactment for the time being in force relating

to income-tax or any other tax on income, profits or gains, the Corporation shall not be liable to pay any income-tax or any other tax in respect of–

(a) any income, profits or gains, accruing or arising out of the Fund of the Corporation or any amount received in that Fund; and

(b) any income, profits or gains, derived or any amount received, by the Corporation.

CHAPTER IV

MISCELLANEOUS

Power of Central Government to Give Directions

23. (1) The Central Government may, from time to time as and when occasion arises, issue to the Corporation such directions as it may think necessary in the interests of the sovereignty, unity and integrity of India or the security of the State or preservation of public order requiring it not to make a broadcast on a matter specified in the direction or to make a broadcast on any matter of public importance specified in the direction.

(2) Where the Corporation makes a broadcast in pursuance of the direction issued under sub-section (1), the fact that such broadcast has been made in pursuance of such direction may also be announced alongwith such broadcast, if the Corporation so desires.

(3) A copy of every direction issued under sub-section (1) shall be laid before each House of Parliament.

Power of Central Government to Obtain Information

24. The Central Government may require the Corporation to furnish such information as that Government may consider necessary.

Report to Parliament in Certain Matters and Recommendations as to Action Against the Board

25. (1) Where the Board persistently makes default in complying with any directions issued under Section 23 or fails to supply the information required under Section 24, the Central Government may prepare a report thereof and lay it before

each House of Parliament for any recommendation thereof as to any action (including supersession of the Board) which may be taken against the Board.

(2) On the recommendation of the Parliament, the President may by notification supersede the Board for such period not exceeding six months as may be specified in the notification :

Provided that before issuing the notification under this sub-section, the President shall give a reasonable opportunity to the Board to show cause as to why it should not be superseded and shall consider the explanation and objections, if any, of the Board.

(3) Upon the publication of the notification under sub-section (2) –

(a) all the Members shall, as from the date of supersession, vacate their offices as such;

(b) all the powers, functions and duties which may, by or under the provisions of this Act be exercised or discharged by or on behalf of the Board, shall, until the Board is reconstituted under this Act, be exercised and discharged by such person or persons as the President may direct.

(4) On the expiration of the period of supersession specified in the notification issued under sub-section (2), the President may reconstitute the Board by fresh appointments, and in such a case any person who had vacated his office under clause (a) of sub-section (3) shall not be disqualified for appointment :

Provided that the President may, at any time before the expiration of the period of supersession, take action under this sub-section.

(5) The Central Government shall cause the notification issued under sub-section (2) and a full report of the action taken under this section to be laid before each House of Parliament.

Office of Member not to Disqualify a Member of Parliament

26. It is hereby declared that the office of the member of the Broadcasting Council or of the Committee constituted under Section 13 shall not disqualify its holder for being chosen as, or for being, a Member of either House of Parliament.

Chairman, Members, etc. to be Public Servants

27. The Chairman and every other Member, every officer or other employee of the Corporation and every member of a Committee thereof, the President and every member of the Broadcasting Council or every member of a Regional Council or a Recruitment Board shall be deemed to be a public servant within the meaning of Section 21 of the Indian Penal Code.

Protection of Action Taken in Good Faith

28. No suit or other legal proceeding shall lie against the Corporation, the Chairman or any Member or officer or other employee thereof or the President or a member of the Broadcasting Council or a member of a Regional Council or a Recruitment Board for anything which is in good faith done or intended to be done in pursuance of this Act or of any rules or regulations made thereunder.

Authentication of Orders and other Instruments of Corporation

29. All orders and decisions of the Corporation shall be authenticated by the signature of the Chairman or any other Member authorised by the Corporation in this behalf and all other instruments executed by the Corporation shall be authenticated by the signature of the Executive Member or by any officer of the Corporation authorised by him in this behalf.

Delegation of Powers

30. The Corporation may, by general or special order, delegate to the Chairman or any other Member or to any officer of the Corporation, subject to such conditions and limitations, if any, as may be specified therein, such of its powers and duties under this Act as it may deem fit.

31. (1) The Corporation shall prepare once in every calendar year, in such form and within such time as may be prescribed, an annual report giving a full account of its activities (including the recommendations and suggestions made by the Broadcasting Council and the action taken thereon) during the previous year and copies thereof shall be forwarded to the Central Government and that Government shall cause the same to be laid before

each House of Parliament.

(2) The Broadcasting Council shall prepare once in every calendar year, in such form and within such time as may be prescribed, an annual report giving a full account of its activities during the previous year and copies thereof shall be forwarded to the Central Government and that Government shall cause the same to be laid before each House of Parliament.

Power to Make Rules

32. (1) The Central Government may, by notification, make rules for carrying out the provisions of this Act.

(2) In particular, and without prejudice to the generality of the foregoing power, such rules may provide for all or any of the following matters, namely :

(a) the salaries and allowances and conditions of service in respect of leave, pension (if any), provident fund and other matters in relation to the whole-time Members under sub-section (7) of Section 6;

(b) the allowances payable to the Chairman and part-time Members under sub-section (8) of Section 6;

(c) the control, restriction and conditions subject to which the Corporation may appoint officers and other employees under sub-section (1) of Section 9;

(d) the manner in which and the conditions and restrictions subject to which a Recruitment Board may be established under sub-section (1) of Section 10;

(e) the qualification and other conditions of service of the members of a Recruitment Board and their period of office under sub-section (2) of Section 10;

(f) the terms and conditions in accordance with which the deputation may be regulated under sub-section (2) of Section 11;

(g) the salary and allowances and conditions of service in respect of leave, pension (if any), provident fund and other matters in relation to the President of the Broadcasting Council under sub-section (5) of Section 14;

(h) the allowances payable to other members of the Broadcasting Council and the members of the Regional Councils, under sub-section (6) of Section 14;

(i) the manner in which the Corporation may invest its moneys under Section 19;

(j) the form and the manner in which the annual statement of accounts shall be prepared under sub-section (1) of Section 21;

(k) the form in which and the time within which, the Corporation and the Broadcasting Council shall prepare their annual report under Section 31;

(l) any other matter which is required to be, or may be, prescribed.

Power to Make Regulations

33. (1) The Corporation may, by notification, make regulations not inconsistent with this Act and the rules made thereunder for enabling it to perform its functions under this Act.

(2) Without prejudice to the generality of the foregoing power such regulations may provide for all or any of the following matters, namely :

(a) the manner in which and the purpose for which the Corporation may associate with itself any person under sub-section (7) of Section 3;

(b) the times and places at which meetings of the Board shall be held and, the procedure to be followed thereat and the quorum necessary for the transaction of the business at a meeting of the Board under sub-section (1) of Section 8;

(c) the methods of recruitment and conditions of service of officers and other employees of the Corporation under sub-section (2) of Section 9;

(d) the remuneration and other conditions of service, including pension, leave and provident fund in relation to an officer or other employee of the Corporation under sub-section (4) of Section 11;

(e) the authority competent to make certain appointments referred to in clause (a) of sub-section (6) of Section 11;

(f) the services which may be provided by the Corporation under clause (f) of sub-section (3) of Section 12;

(g) the determination and levy of fees and other service charges in respect of advertisement and other programmes under sub-section (7) of Section 12;

(h) the manner in which and the period within which complaints may be made under sub-section (2) of Section 15;

(i) any other matter in respect of which provision is, in the opinion of the Corporation, necessary for the performance of its functions under this Act :

Provided that the regulations under clause (c) or clause (d) shall be made only with the prior approval of the Central Government.

Rules and Regulations to be Laid before Parliament

34. Every rule and every regulation made under this Act shall be laid, as soon as may be after it is made, before each House of Parliament, while it is in session for a total period of thirty days which may be comprised in one session or in two or more successive sessions, and if, before the expiry of the session immediately following the session or the successive sessions aforesaid, both Houses agree in making any modification in the rule or regulation, or both Houses agree that the rule or regulation should not be made, the rule or regulation shall thereafter have effect only in such modified form or be of no effect, as the case may be; so, however, that any such modification or annulment shall be without prejudice to the validity of anything previously done under that rule or regulation.

Power to Remove Difficulties

35. If any difficulty arises in giving effect to the provisions of this Act, the Central Government may, by order, published in the Official Gazette, make such provisions, not inconsistent with the provisions of this Act, as it may deem necessary, for the removal of the difficulty :

Provided that no such order shall be made after the expiry of a period of three years from the appointed day.

ANNEXURE-II

MEDIA POLICY FOR COMMUNITY RADIO BANGALORE DECLARATION SEPTEMBER 1996

The present decade is significant for the country's development in a number of ways. Decentralisation of governance, professed emphasis on social services, impact of the communication revolution and increased general awareness of human rights have opened up opportunities and challenges on an unprecedented scale. If these far-reaching changes in the economic, technological and social spheres can be harnessed to empower the impoverished and disadvantaged millions, the process of development can be given a fresh impetus and a more purposive orientation.

Information is a key element in socio-economic development. The capacity to communicate and have access to the media greatly influences information use and its impact. People will also generate new and vital information when the media are democratised. Therefore, the need for a dynamic and people-oriented policy on communication cannot be over-stressed. A variety of factors peculiar to the Indian situation today suggest that a liberalised policy on radio broadcasting with emphasis on community participation and local control, will be appropriate to the current development strategy. Formulation of broadcasting policy in respect of radio should aim at optimum utilisation of airwaves in the public interest, facilitating community management of information exchange and participatory modes of decision-making. Limited experiments in this direction by the All India Radio through the so-called Local Radio Stations have demonstrated the immense benefits and potential of this approach. It is proposed here to institutionalise and expand the concept of community broadcasting, with the involvement of people at all levels across the length and breadth of this vast country. Democracy and good governance make such a change imperative; and it is an essential pre-requisite for better enjoyment of human rights.

More than sixty persons representing All India Radio, universities, non-government organisations involved in development activities, journalists, and members of the broadcasting establishment, discussed the problems and prospects for evolving a sectoral broadcasting policy during a four-day Consultation on Community Radio and Media Policy held in Bangalore during September 11-14, 1996. The discussions focused on the means and modalities of establishing, at the earliest, a democratic system of public service broadcasting in the country. The Consultation resulted in a consensus statement, here called the Bangalore Declaration on Radio, which outlines the basic elements of a desirable media policy on radio broadcasting.

Given the set of current policies aimed at decentralisation, liberalisation and people's participation in several areas, it is hoped that the Bangalore Declaration on Radio will receive the timely and earnest attention of relevant authorities, and that the concept of community radio will soon become a reality. Meanwhile, it is important that universities, public bodies, women's organisations, dalit associations to name a few–prepare themselves for the task of establishment and participatory management of community radio stations with a development agenda appropriate to, and determined by, their communities : this is the near-term challenge and unique opportunity. Community radio will bring fresh validity and meaning to the concept of decentralised governance envisaged by 73rd and 74th Amendments of the Constitution; and catalyse the process of people's empowerment through information and education. It is with this hope and reasonable expectation that the Bangalore Declaration is being presented to the public for a wider discussion, support and follow-up action.

NATIONAL BROADCASTING POLICY : TOWARDS PUBLIC SERVICE BROADCASTING THROUGH COMMUNITY RADIO

Preamble

Whereas the right to impart and receive information is recognised as a guaranteed basic right under the Constitution; and

Whereas, the Supreme Court of India has declared airwaves as public property to be utilised for promoting public good and ventilating plurality and diversity of views, opinions and ideas, and that towards this end, the broadcasting media should be under the control of the public as distinct from the Government; and

Whereas, it is necessary to promote local broadcasting through community-based radio stations for the successful implementation of decentralised governance envisaged under the 73rd and 74th Constitutional Amendments; and

Whereas, successive committees appointed by the Government to examine the issue have recommended the adoption of a comprehensive national broadcasting policy with participation of the private sector, NGOs, local self-government institutions and other community groups providing a framework for media development in the context of technological progress and national needs :

It is now proposed to present the elements of a National Broadcasting Policy on the basis of which the Parliament could enact legislation for regulating the airwaves as a public property, whose utilisation in the context of electronic media–TV and radio–is intended to promote public good.

[Since the Bangalore Consultation was specifically focused on radio broadcasting, the statement of policy contained herein is limited to only the radio broadcasting sector, and in particular, its community broadcasting application. It is our hope that this statement would be included appropriately in the National Media Policy which would cover all media–Television, Radio, Cinema and Print.]

Statement

The Legislative Imperative

1. Airwaves are public property and must be used for public good. Public good is not served optimally when there is a monopoly by the government over this public resource or when liberalisation of broadcasting is confined to commercial use of airwaves. Access to the airwaves by everyone who acts in the public interest is the *sine qua non* of public good in this context. This is also part of the right to information, a guaranteed constitutional right.

Need for Community Radio

2. Centralised one-way broadcasting at various levels of aggregation has limited scope to serve the goals of development, especially in the context of pluralism and diversity which is a

singular characteristic of Indian society. As such the regulatory framework should promote a decentralised system of radio broadcasting.

3. Community broadcasting is a concept relevant to social cohesion, development for conviviality and national integration. Community radio is public service broadcasting in its most decentralised and its most democratic form. A community radio station serves a defined geographical area of a village or groups of villages, and is owned and managed by organisations serving a given community.
4. A community radio station would, besides educating and entertaining people, connect people with people through participatory or circular communication, connect with organisations and communities, and finally, connect people with government and public service agencies. These needs are not met under the current framework.
5. In a number of ways, community or public interest broadcasting can be termed Equal Opportunity Broadcasting, which is essential not only for effective democratisation of a public resource viz. the airwaves, but also in the context of our plural society with a multiplicity of languages, cultures and ethnic groups.

Policy Formulation

6. Radio combines the benefits of low cost, and wide reach and access. When used in a community setting with limited area coverage, for example as in FM radio broadcasting, it offers many exciting possibilities for fulfilling the developmental goals and aspirations of the people, and wider choices in accessing information from diverse sources within and outside the community. Its potential for creating social change has been demonstrated in many parts of the world. For these reasons, control of community broadcasting should be vested with the community rather than with the government or private commercial enterprises.
7. The present centralised structure of broadcasting is not conducive to people-centred, participatory methods of communication. The potential of the medium to promote community development is largely unrealised. The involvement of other public bodies, NGOs, professional associations, etc. in the utilisation and management of airwaves is essential for

realising this potential to promote the welfare of millions in communities, and therefore the public good. The need for structural changes in airwaves management in order to accommodate localised initiatives is a logical result of this imperative.

8. As recommended by the Paswan Committee (1996), while there can be a single national policy which addresses macro-level issues common to all media, medium-specific policies are required for each medium, taking into account the coverage, reach, cost structure, technology, administration and, finally, the social application of the medium.
9. Radio broadcasting, like other media, has developed around power centres and power structures in society, resulting in disparity of access and use among different sections of society. Changes in media policy which seek to mitigate, if not eliminate, these disparities require political will and a people-centred, bottom-up approach already contemplated in the 74th Constitutional Amendment. Community radio is an illustration of this evolutionary process; therefore, it needs to be acknowledged and supported.
10. At the operational level, a regulatory structure which is independent of the government and existing broadcasting organisations, public and private, should be set up in order to facilitate and support all licensed broadcasters in the country and their shared use of the airwaves under a fair and reasonable regime, consonant with national interests and priorities. The structure must address a variety of issues such as : licensing criteria for various categories of broadcasting, technical and service standards, technical support and training, social and administrative auditing, funding sources and support, etc.
11. Frequency spectrum resources–particularly in the band allocated for FM radio broadcasting viz. 86-108 MHz–will be made available on a shared basis with existing services for community broadcasting applications. Efforts will be made to evolve a national plan for the allocation of frequencies in this band so that the establishment of a large number of community radio stations across the country, based on a high degree of frequency re-use, is facilitated.

Note

Far-reaching structural and managerial changes in the regulation and administration of the electronic media by the Government have been recommended by the various committees appointed to examine the issue. These have been partly reflected in the Prasar Bharati Act (1990). This Act might require changes if the statement on policy contained in the Bangalore Declaration is to be adopted in legislation. At the Bangalore Consultation on Community Radio and Media Policy, a possible structure for autonomous public service radio broadcasting was discussed, taking into account the appropriate features of such a structure and also the ground realities of the country. The outcome of these discussions, titled "Structure of Public Service Radio Broadcasting in India; Some Essential Features," is appended to this document as Annexure I. Admittedly, some of the features, listed therein–particularly those related to funding, organisation, and inter-linkages–need to be elaborated in more detail, after further discussion. However, the proposed structure provides a basis for discussion, once the important policy decisions on public control of airwaves and the concept of community broadcasting are taken.

ESSENTIAL FEATURES OF A PUBLIC SERVICE RADIO BROADCASTING STRUCTURE

Regulatory Authority and Licensing Criteria

1. To regulate and oversee broadcasting activities in the country under a liberalised framework, a National Broadcast Trust (NBT) as outlined by the Verghese Committee (1978) should be established. The NBT will be an autonomous body made up of eminent professionals and other individuals, and would be free from government control.
2. The current framework adopted by national broadcasting agencies is of three-tier broadcasting viz., national, regional and local. It is proposed here to enable the setting up of a separate and independent tier at village/community level. The nature of community broadcasting warrants that it should be owned and managed by the community itself. For all other purposes, community broadcasting will be entitled to the same rights and obligations as broadcasting agencies at the other levels.
3. The term 'community' for this purpose shall mean a non-rigid group of people living in a bounded geographical area which is determined by the reach of the radio broadcasting station, and

who share a commonality of concerns, interests and aspirations.

4. No monopoly or exclusive control by any interest group within the community will be permitted. The community, in all its diversity, should exercise democratic control over community broadcasting and establish appropriate systems of accountability, within and outside the community, as may be necessary. Equal opportunity to all groups in the community in respect of access to communication will be the distinct feature of a community radio station.
5. Given the complexity of local communities in India, the regulatory authority may grant licences to other bodies serving the public interest, in a phased manner. In the first phase, universities with extension activities (for example, agricultural universities, medical institutions, adult and legal literary organisations), registered cooperatives, women's cooperatives and autonomous public bodies fulfilling minimum criteria for eligibility should be granted licences to set up community radio stations to serve the public interest.

Programming

6. While granting licences for community radio stations, the NBT will direct the All India Radio to provide the required development and technical support to the licensees, including training of technicians and programmers. The licensee shall bear the costs of programming as well as the operating and maintenance costs of the community radio station.
7. A local advisroy committee, representing varied public interests within the community, shall be appointed, preferably by a local judge, for a fixed tenure. The committee shall guide the programme content of transmissions from the community radio station. The committee. shall also guide all other collective decisions relating to the operations of the community radio station, such as hours of transmission, scheduling and so on.
8. The interactive format for programming shall be encouraged to make the programmes truly participatory. A baseline programming element will relate to environment, health, women's empowerment, education, legal matters and other such public interest issues.
9. Building linkages between private broadcasters, on the one hand, and local self-government institutions, educational and professional bodies, and special interest groups on the other

hand, will be essential. It is important that these groups are provided access to the medium in order to serve the public interest in all its dimensions. Such participation would of course be innovative. Community radio stations should also work in concert with government agencies in reaching out to and promoting the economic and social welfare of the marginalised and disadvantaged groups within the community.

10. The NBT may lay down guidelines on programme content promotive of public interest and may monitor community radio stations. It may exercise sanctions to discipline erring stations, again in the public interest as elucidated in the policy statement.
11. Community radio stations may be obliged to broadcast during an agreed period, national messages provided by the AIR, for which the radio networking receiver facilities may be extended by the latter.

Role of All-India Radio

12. All-India Radio, with its extensive infrastructure, trained personnel and vast experience, shall play a crucial but supportive role in the development of community radio broadcasting, under the guidance of the NBT. Its new role in this sphere shall be clearly defined and the necessary organisational restructuring of the AIR shall be carried out to help carry out its new role effectively.
13. All-India Radio shall provide assistance to licenced community broadcasters in the design of their radio stations in respect of intended coverage, technical configuration of transmitter and studio, and standardisation of the service at low investment levels; it shall also provide technical training for operation of the facilities. The AIR could also evolve a Handbook/Manual on community broadcasting, based on technical/system studies, which can be a source-book for the NBT in decision-making on technical matters and for potential broadcasters.

Checks and Balances

14. To ensure social and public accountability, it is desirable that each community radio station evolves its own code of conduct and strictly abide by it in its programming, transmissions and management. A local Ombudsman consisting of three persons who are widely respected for their experience and integrity

shall be attached to each radio station. The Ombudsman will entertain complaints from individuals and institutions and can decide on the culpability or otherwise of the radio station concerned. Once indicted by the Ombudsman, the radio station will have to make amends and strictly abide by its directions. More than three such violations may entail withdrawal of the licence, or other privileges of the station or its officers.

15. The station can also be brought under the Consumer Protection Act as a provider of services in a limited number of cases.
16. If the regulatory authority directs a radio station to close down or if the station cannot be run for other reasons, including emergencies, AIR shall have the authority to keep the assets in trust till the problem is resolved and the station resumes broadcasting. If AIR is to finally and irrevocably take over the station alongwith the assets, the owners shall receive due compensation in respect of the assets taken over by AIR.

Funding and Management

17. All community radio stations will work on the principle of no-profit, no-loss. Initial capital expenditure shall be met largely by a grant through the National Broadcast Trust, contributions from member institutions, donations from the public, fees from listeners for broadcast of local messages and advertisements, sponsorship charges, radio club subscriptions, etc.
18. The appropriate legal form for a community radio station could be either a Society registered under the Societies Registration Act; or a Trust under the Trusts Act. Other options for management of community radio stations could also be examined by the regulatory authority and a final organisational format can be arrived at.

Interim Measures by the Government

19. In order to promote community broadcasing and the development of necessary skills in programming and other areas in the interim period when the legislative framework of community radio and follow-up actions are pending, the Government could provide for the following :
    (a) Airtime in AIR Local Radio Stations for programmes developed by different community and special interest groups to provide regular community interaction with

audiences within the coverage areas.

(b) Airtime in Private Radio Stations licensed by the Government for institutions and public bodies within the community to provide regular community programming to audiences within the coverage areas.

(c) Mandatory setting up of Community Broadcasting Stations areas by licensed broadcasters–government and private–as a fraction of the total number of local radio stations licence by the NBT e.g. for every five Local Radio Station licences granted, the licencee is obliged to set up one community radio station.

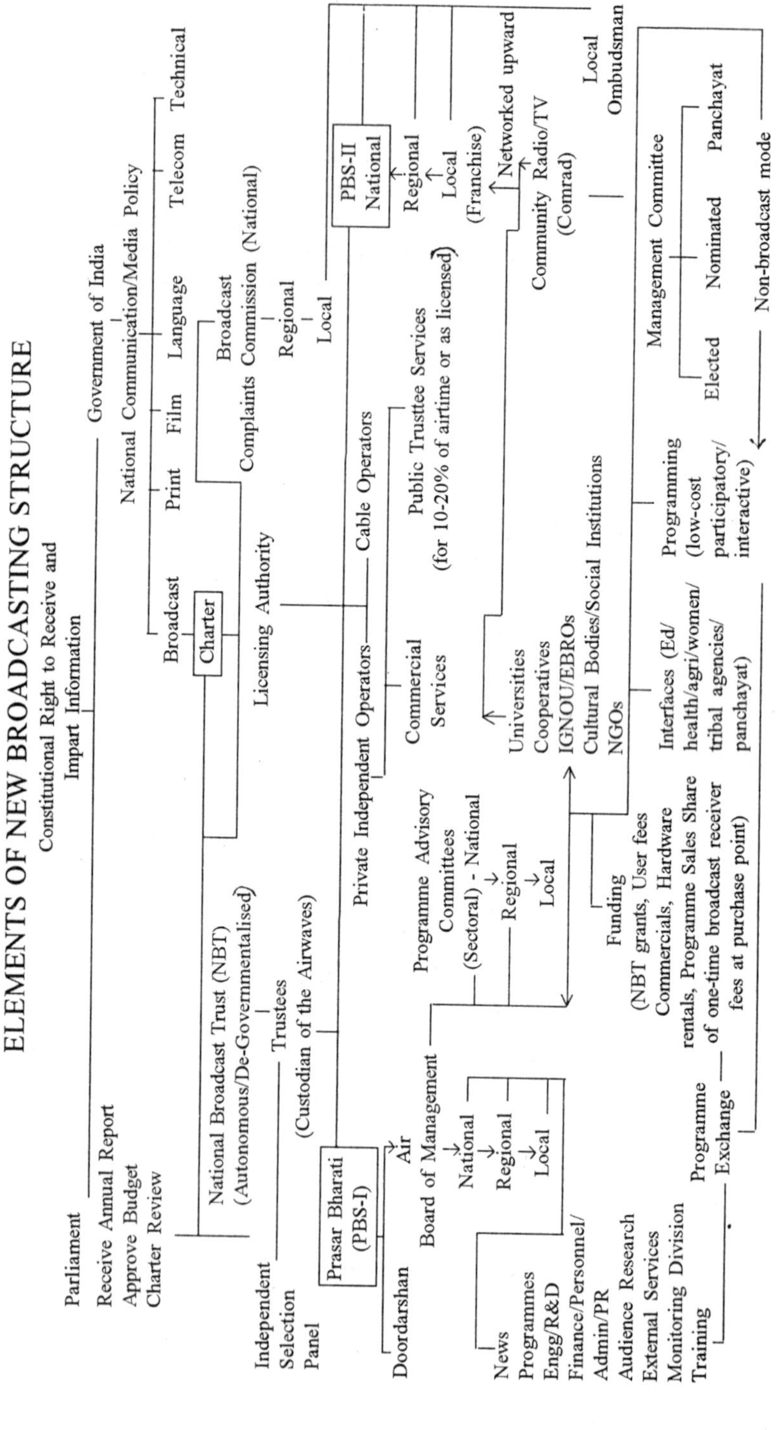
ELEMENTS OF NEW BROADCASTING STRUCTURE
Constitutional Right to Receive and Impart Information
Parliament
Receive Annual Report
Approve Budget
Charter Review
Government of India
National Communication/Media Policy
Broadcast
Print
Film
Language
Telecom
Technical
Charter
National Broadcast Trust (NBT)
(Autonomous/De-Governmentalised)
Broadcast Complaints Commission (National)
Regional
Local
Independent Selection Panel
Trustees
(Custodian of the Airwaves)
Licensing Authority
Private Independent Operators
Cable Operators
Prasar Bharati (PBS-I)
Doordarshan
Air
PBS-II National
Regional
Local
(Franchise)
Networked upward
Community Radio/TV (Comrad)
Local Ombudsman
Board of Management
National
Regional
Local
Programme Advisory Committees (Sectoral) - National
Regional
Local
Commercial Services
Public Trusttee Services (for 10-20% of airtime or as licensed)
Universities
Cooperatives
IGNOU/EBROs
Cultural Bodies/Social Institutions
NGOs
News
Programmes
Engg/R&D
Finance/Personnel/
Admin/PR
Audience Research
External Services
Monitoring Division
Training
Programme Exchange
Funding
(NBT grants, User fees Commercials, Hardware rentals, Programme Sales Share of one-time broadcast receiver fees at purchase point)
Interfaces (Ed/health/agri/women/tribal agencies/panchayat)
Programming (low-cost participatory/interactive)
Management Committee
Elected
Nominated
Panchayat
Non-broadcast mode

ANNEXURE-III

RIGHT TO INFORMATION: TOWARDS A NEW FREEDOM

The common minimum programme, announced by the United Front Government, on its assumption of office, included a commitment to put on the Statute book people's Right to Information. This has excited popular imagination as nothing else has, since the enactment of Article-19-1(A) of the Constitution, guaranteeing a citizen's right to freedom of speech and expression. The right to information will complete the process that began with the grant of freedom of speech and expression.

A draft bill on the Right to Information has been drawn up by the Press Council of India, following a Workshop on the subject organised by the Council in August 1996. It has been sent to the Union Government for consideration. The Workshop was attended by eminent parliamentarians, jurists, journalists, academics and political activists.

The common refrain of the Workshop was that the proposed right will go a long way towards ensuring transparency in the working of the government which will lift the country from the depths of corruption and inefficiency it has sunk to.

It was felt that the demand for right to information was basic to the successful functioning of any democracy. If the people had a right to governance, they cannot exercise it effectively without information.

The draft bill, seeks to ensure that every public authority shall be under a duty to maintain all its records properly and to make it available to any person seeking information. Nothing shall preclude the right of any person to obtain such information as would affect his life or liberty.

The draft bill, however, provides that information, disclosure of which can affect the sovereignty and integrity of the country, can be withheld by the authority concerned.

RIGHT TO INFORMATION DRAFT BILL

A Bill to make provision for securing right to information and for matters connected therewith or incidental thereto.

Be it enacted by Parliament in the forty-seventh year of the Republic of India as follows :

1. (1) This Act may be called the Right to Information Act, 1996.
   (2) It extends to the whole of India, except the State of Jammu and Kashmir.
   (3) It shall come into force on such date as the Central Government may, by notification in the Official Gazette, appoint.

Definitions

2. In this Act, unless the context otherwise requires :

(a) "Information" means any material or information relating to the affairs, administration or conduct of a public authority in respect of which public has a right to know, and includes any document or record relating to the affairs of the Public Authority.

(b) "prescribed" means prescribed by rules under the Act;

(c) "Public authority" includes–

   (i) the Government and Parliament of India and the Government and Legislature of each of the States and of local or other authorities within the territory of India or under the control of the Government of India; and

   (ii) a company, corporation, trust, firm, society or a cooperative society, whether owned or controlled by the Government or by private individuals and institutions whose activities affect the public interest;

   The expressions company, corporation, trust, firm, society and cooperative society shall have the same meaning as assigned to them in the respective Acts under which they are registered.

(d) "right to information" means the right of access to information and includes the inspection, taking notes and extracts and obtaining certified copies of documents or records of any public authority;

(e) "State Government" includes the administrator of a Union Territory appointed by the President under article 239 of the Constitution.

Right to Information

3. (1) Every citizen shall have a right to information from a public authority.

(2) Every public authority shall be under a duty to maintain all its rĕcords duly catalogued and indexed and subject to the provisions of Section 4, to make available to any person requesting information from it, the information sought for, including such information as it is under an obligation to obtain and furnish and shall not withhold any information or limit its availability.

(3) Nothing in the foregoing provisions of the section shall preclude the right of any person to obtain such information as would affect his life or liberty.

(4) It shall be mandatory for each custodial establishment such as a police lock-up, jail, mental asylum, remand house, women's homes, baggars' homes, etc. to appoint a visitors' committee comprising independent citizens which shall have full access to them at all hours of the day and night and to their records and inmates.

Restrictions on Right to Information

4. The public authority may, for reasons to be recorded in writing, withhold:

(a) information, the disclosure or contents of which will prejudicially affect the sovereignty and integrity of India, security of the State and friendly relations with foreign states, public order, investigation of an offence or which leads to incitement to an offence;

(b) information relating to personal or other information, the disclosure of which has no relationship to any public activity or to which the public has no interest and would constitute a clear and unwarranted invasion of personal privacy; and

(c) trade and commercial secrets protected by law ;

Provided that information which cannot be denied to Parliament or State Legislature shall not be denied to any citizen.

Procedure for the Supply of Information

5. (1) A person desiring information from a public authority under this Act shall make an application in such manner, as may be prescribed to the person for the time being in charge of the public authority.
   (2) The information asked for under sub-section (1) shall be furnished to the applicant by the public authority within thirty days of the application under sub-section (1), and where such information relates to the life or liberty of the individual, the information shall be supplied within forty-eight hours of the making of the application.
   (3) The information asked for under sub-section (1) shall be supplied in writing either in English or Hindi or in the language of the state in which the public authority is located.
   (4) Every public authority shall have a right to charge any fees for the supply of any document subject to the condition that such fees shall not exceed the cost of making and supplying the document.

6. If the information asked for under sub-section (1) of Section (4) is refused by any public authority, the reasons for such refusal shall be given by such authority in writing.

Appeal

7. An appeal against a refusal to furnish any document by a public authority under this Act shall lie to the Principal Civil Judge of ordinary original civil jurisdiction of the area in which the office of the public authority is located and any appeal shall be preferred within fifteen days of the date of refusal of the supply of information by the public authority and any appeal so preferred shall be disposed of by the Civil Judge within thirty days of its filing, by a written order and giving reasons for its decision.

Penalties

8. (1) Any individual responsible for providing any information under this Act shall be personally liable for furnishing the information within the period specified therein.
   (2) Where the information asked for under this Act is not furnished within the time specified, the person responsible for supplying

the information shall be punishable with a fine of rupees fifty for every day of delay after the day by which it is required to be supplied.

(3) If any person responsible for providing any information under this Act, fails to furnish such information or furnishes any information which is false in any material particular and which he knows or has reasonable cause to believe to be false or does not believe it to be true, he shall be punishable with fine which shall not be less than rupees one thousand and which may extend to rupees five thousand :

Provided that the court may, for reasons to be recorded in writing levy a fine of less than one thousand rupees.

Offences by Companies

9. (1) Where an offence under this Act has been committed by a company, every person who, at the time the offence was committed, was in charge of, and was responsible, to the company for the conduct of the business of the company, shall be deemed to be guilty of the offence and shall be liable to be proceeded against and punished accordingly :

Provided that nothing contained in this section shall render any such person liable to any punishment provided in this Act if he proves that the offence was committed without his knowledge or that he had exercised all due diligence to prevent the commission of such offence;

(2) Notwithstanding anything contained in sub-section (1) where an offence under this Act has been committeed by a company and it is proved that the offence has been committed with the consent or connivance of or is attributable to any neglect on the part of any director, manager, secretary or other officer of the company, such director, manager, secretary or officer shall also be deemed to be guilty of that offence and shall be liable to be proceeded against and punished accordingly.

Explanation : For the purposes of this section :

(a) "company" means any body corporate, and includes a firm or association of individuals, and

(b) "director" in relation to a firm means a partner of the firm.

Offences to be Cognizable

10. Notwithstanding anything contained in the Code of Criminal Procedure 1973, every offence under this Act shall be cognizable.
11. Notwithstanding anything contained in this Act or in any other law for the time being in force, it shall be the duty of every public authority which proposes to initiate any project or activity, shall, subject to the provisions of Section 4, notify, as soon as possible, and in any case well before the commencement of the project of activity, for the information of the general public and the persons to be affected, such matters within its knowledge and control as affect the general public or are vital to life, health and livelihood.

Protection of Action Taken in Good Faith

12. No suit, prosecution or other legal protection proceeding shall lie against the Central Government or a State Government or any individual, for anything which is in good faith done or intended to be done under the provisions of this Act or any rule made thereunder.
13. The provisions of this Act shall have effect notwithstanding anything inconsistent therewith contained in any other law for the time being in force other than this Act.

Power of the Central Government to Make Rules

14. (1) The Central Government may, by notification in the Official Gazette, make rules to carry out the provisions of this Act.
    (2) Without prejudice to the generality of the foregoing power, such rules may provide for all matters in respect of which rules are required to be made by the Central Government under this Act.

Power of State Government to Make Rules

15. A State Government may, by notification in the Official Gazette, make rules to provide for any matter in respect of which rules are required to be made by the State Government under this Act.

Laying of Rules

16. (1) Every rule made under this Act by the Central Government shall

be laid, as soon as may be after it is made, before each House of Parliament, while it is in session, for a total period of thirty days which may be comprised in one session or in two or more successive sessions and if, before the expiry of the session immediately following the session or the successive sessions aforesaid, both Houses agree that the rule should not be made, the rule shall thereafter have effect only in such modified form or be of no effect, as the case may be; so, however that any such modification or annulment shall be without prejudice to the validity of anything previously done under that rule.

(2) Every rule made under this Act by a State Government shall be laid, as soon as may be after it is made, before the state Legislature.

Annexure III has been excerpted from the Mediascape' 97, published by the National Media Centre on the occasion of the Asian Media Convention, 22-24 February 1997.

LIST OF ABBREVIATIONS

AMIC Asian Mass Communication Research and Information Centre/Asian Media Information and Communication Centre
APIBD Asian Pacific Institute of Broadcasting Development
BES Broadcasting Engineering Society
C-DOT Centre for Development of Telematics
CNN Cable News Network
COL Commonwealth of Learning
DAB Digital Audio Broadcasting
DPAP Drought Prone Area Programme
DSC Development Support Communication
ERNET Education and Research in Network
FES Friedrich Ebert Stiftung
GDP Gross Development Product
GNP Gross National Product
HDI Human Development Index
HYV High Yielding Variety
IARI Indian Agricultural Research Institute
ICDP International Conference on Population and Development
IGNOU Indira Gandhi National Open University
INSAT Indian National Satellite
IPDC International Programme for Development Communication
IRDP Integrated Rural Development Programme
IRPA Indian Rural Press Association
ISRO Indian Space Research Organisation
MFALA Marginal Farmer and Agriculture Labour Agency
NGO Non-Governmental Organisation
NIMME National Institute of Multi-Media Education
NICNET National Information Centre Network

NPG	National Performance Gap
ORT	Oral Rehydration Therapy
RAX	Rural Automatic Exchange
RNI	Registrar of Newspapers of India
SAC	Sapce Application Centre
SFDA	Small Farmers Development Agency
SITE	Satellite Instructional Television Experiment
UNDP	United Nations Development Programme
UNESCO	United Nations Educational Scientific and Cultural Organization
VOU	Virtual Online University

BIBLIOGRAPHY

Agarwal, Bina (1994). *A Field of One's Own, Gender and Land Rights in South Asia.* Cambridge University Press.

Agnihotri, V. K. (ed.) (1995). *Public Policy, Analysis and Design.* Concept Publishing Company, New Delhi.

Akash Bharati : National Broadcast Trust (1978)–Report of the Working Group on Autonomy for Akashvani and Doordarshan. Ministry of Information and Broadcasting, New Delhi.

An Indian Personality for Television (1985). Report of the Working Group on Software for Doordarshan. Ministry of Information and Broadcasting, New Delhi.

Asian Mass Communication Research and Information Centre (1985). *Satellite Technology. The Communication Equaliser,* Singapore.

Asian Mass Communication Research and Information Centre (1993). *Role of Media in a National Crisis,* Singapore.

Asian Mass Communication Research and Information Centre (1996). *Opening Windows–Issues in Communication,* Singapore.

Bhargava, G. S. (ed) (1991). *Government Media–Autonomy and After.* Concept Publishing Company, New Delhi.

Bhatt, Ela R. (1988). *Shramshakti.* National Commission on Self-Employed Women and Women in Informal Sector, New Delhi.

Chandra, Bipan (1993). *Essays on Contemporary India.* Har-Anand Publications, New Delhi.

Dreze, Jean and Sen, Amartya (1995). *India : Economic Development and Social Opportunity.* Oxford University Press, Oxford and New Delhi.

Eapen, K. E. (ed.) (1989). Role of Radio in Growth and Development. Mass Media Education, Research and Information Centre, Bangalore.

Goonasekera, Anura (1993). Media Technology and Social Imperatives: An Examination of Communication Policies in Selected Asian Countries. *Media Asia* 20 (4). AMIC Singapore.

This bibliography contains a list of only those sources which have been actually used in the book, and not the many others which have been consulted by the author but not specifically used. Minor references in the text, such as references to newspaper, have been omitted.

Gupta, V. S. (1995). *Third Revolution in Indian Perspective–Contemporary Issues and Themes in Communication.* Concept Publishing Company, New Delhi.

Gupta, V. S. and Aggarwal, Vir Bala (1996). *Media Policy and Nation Building, Select Issues and Themes.* Concept Publishing Company, New Delhi.

Gupta, V. S. (1996). *Towards a Rural Channel.* Voices IV (1), Bangalore.

Gupta, V. S. (1996). Media Strategies for Rural Development. *Kurukshetra* XLIV (4-5). Publications Division, New Delhi.

Gupta, V. S. and Dyal, Rajeshwar (eds.) (1996). *National Media Policy.* Concept Publishing Company, New Delhi.

Human Rights–The New Consensus (1994). Regency Press, Golden House, London.

Independent Commission on Population and Quality of life (1996). *Caring for the Future.* Oxford University Press, Oxford.

Joseph, Ammu and Sharma, Kalpana (1994). *Whose News? The Media and Women's Issues.* Sage Publications, New Delhi.

Joshi, P. C. (1986). *Communication and Nation-Building : Perspective and Policy.* Sardar Patel Memorial Lectures, Publications Division, New Delhi.

Lee, John A. R. (1976). *Towards Realistic Communication Policies–Recent Trends and Ideas.* Compiled and Analysed, UNESCO, Paris.

Martin William, J. (1995). *The Global Information Society.* Hampshire, Asliba Gower.

McManus, John H. (1996). *Market–driven Journalism–Let the Citizen Beware?* Sage Publications, London.

Masani, Mehra (1976). *Broadcasting and the People.* National Book Trust, New Delhi.

Melkote, Srinivas R. (1991). *Communication for Development in the Third World–Theory and Practice.* Sage Publications, New Delhi.

Ministry of Information & Broadcasting (1996). *National Media Policy–A Working Paper.*

Mitra, Anand (1993). *Television and Popular Culture in India—A study of the Mahabharat.* Sage Publications, New Delhi.

Mody, Bella (1991). *Designing Messages for Development Communication.* Sage Publications, New Delhi.

Mowlana, Hamid and Laurie, Wilson J. (1988). *Communication Technology and Development.* Reports and Papers on Mass Communication, UNESCO, Paris.

Nair, K. S. and White, S. A. (1993). *Perspectives on Development*

Communication. Sage Publications, New Delhi.

Narayanan, V. N. and Sabharwal, Jyoti (1997). *India at 50, Bliss of Hope and Burden of Reality.* Sterling Publishers, New Delhi.

National Media Centre (1997). *Mediascape' 97,* New Delhi.

Nihal Singh, S. (1992). *Your Slip is Showing–Indian Press Today.* UBS Publishing, New Delhi.

Rao, U. R. (1996). *Space Technology for Sustainable Development.* Tata McGraw Hill Publishing Company, New Delhi.

Sarkaria R. S. (1995). A Guide to Journalistic Ethics. *The Hindustan Times,* New Delhi.

Sawant, B. P., Mohan, S. and Reddy, B. P. Jeevan (1995). Supreme Court Judgement in Civil Appeals Nos. 1429-30. Secretary, Ministry of Information and Broadcasting vs. Cricket Association of Bengal.

Shiva Vandana (1991). *Ecology and the Politics of Survival–Conflicts over Natural Resources in India.* Sage Publications, New Delhi.

Singhal, Arvind and Rogers, Everett M. (1989). *India's Information Revolution.* Sage Publications, New Delhi.

Swaminathan, M. S. (1973). *Our Agricultural Future.* Sardar Patel Memorial Lectures, All-India Radio, New Delhi.

Teheranian Majid, Hakimzadeh, Farhad and Vidale, Marcello L. (eds.) (1977). *Communication Policy for National Development.* Routledge and Kegan Paul, London.

Toffler, Alvin (1980). *The Third Wave.* William Collins, London.

UNDP (1993). *Human Development Report.* Oxford University Press, Oxford.

UNDP (1996). *Human Development Report.* Oxford University Press, Oxford.

UNESCO (1982). *Many Voices, One World.* The International Commission on the Study of Communication Problems, Paris.

UNESCO (1997). *World Information Report* 1997-98, Paris.

UNESCO (1997). *Public Service Broadcasting : The Challenge of the Twenty-first Century,* Paris.

Unnikrishanan, Namita and Bajpai, Shailaja (1996). *The Impact of Television Advertising on Children.* Sage Publications, New Delhi.

Venkateswaran, K. S. (1993). *Mass Media, Laws and Regulations in India.* AMIC, Singapore.

Venkateswaran, Sandhya (1995). *Environment, Development and the Gender Gap.* Sage Publications, New Delhi.

Voices (1997). *Media Policy and Community Radio,* Bangalore.

Voices (1997). *Telephone for All : Privilege or Basic Right?* Bangalore.

World Bank (1996). *From Plan to Market.* Oxford University Press, Oxford.

INDEX